U.S. Fire Administration

Many Faces, One Purpose

A Manager's Handbook on Women in Firefighting

FA-196/September 1999

Homeland Security

Prepared by:
Women in the Fire Service
P.O. Box 5446
Madison, Wisconsin 53705
608/233-4768
608/233-4879 fax

Researchers & writers:
Brenda Berkman
Teresa M. Floren
Linda F. Willing

With assistance from:

Debra H. Amesqua	Kim Delgaudio	Carol Pranka
Freda Bailey-Murray	Patricia Doler	Andrea Walter
Joette Borzik	Julia Luckey	Grace Yamane

U.S. Fire Administration

Mission Statement

As an entity of the Department of Homeland Security, the mission of the USFA is to reduce life and economic losses due to fire and related emergencies, through leadership, advocacy, coordination, and support. We serve the Nation independently, in coordination with other Federal agencies, and in partnership with fire protection and emergency service communities. With a commitment to excellence, we provide public education, training, technology, and data initiatives.

Introduction

When an organization moves away from a generations-long tradition of being all-male toward a future that includes men and women equally, a significant change takes place. Change can be upsetting and threatening to those who are used to, and invested in, the way things "have always been."

Fire may know no gender, but people do, and the fire chief of the 1990's spends more time managing people than controlling fire. Increasing numbers of women are becoming firefighters and fire officers, entering and advancing in a field that is still heavily male by both population and tradition. Fire service leaders who are not prepared to manage these workforce changes may find their workforce is managing them instead.

Fire chiefs used to feel comfortably progressive saying, "We'll hire anybody who meets our standards." But the underlying premise to that statement has now been challenged. How have those standards been set? Are they reasonable? Can all current members of the department meet them? What happens to someone who does meet the standards, but then faces a wall of hostility from their new coworkers? What support systems are available for those who are excluded from the privileges of the dominant group?

What fire service leaders have learned, as first men of color and then women of color and white women have entered the field, is that policies that appear to be neutral, or policies that seem to apply equally to everyone, do not necessarily create equal opportunity. Altering the identity of people in a fundamentally unaltered workplace leaves the door open to friction, miscommunication and a host of interpersonal issues that can result in poor performance and a loss of teamwork, or worse. The commitment to equal opportunity in employment has evolved into a commitment to much more than that: to a workforce where diversity itself is valued.

This book was developed to help the fire service leader manage the changing fire service workforce as it becomes progressively more inclusive of larger numbers of women in all ranks. It offers guidance and suggestions from people with experience and expertise, and provides choices and options more often than single "right" answers. The authors hope they have created a guide that, in calling on a wide range of resources, can be useful to a people with a wide range of needs.

As a resource to fire service women, a companion document to *Many Faces, One Purpose* has been created, providing information for women firefighters and women considering firefighting careers. Copies of *Many Women Strong: A Handbook for Women Firefighters* are available from the U.S. Fire Administration (USFA).

Two specific limitations of this document must be noted. While the material has received legal review, it does not consider the requirements of most State or local laws and regulations. And, while the information here was current as of mid-1996, legal issues are always subject to change. This handbook is not intended to be a substitute for competent legal advice, which always should be sought before implementing any policy with legal implications.

The ideas and resources in the handbook have been drawn from the experiences of fire departments across the country in more than two decades of women's involvement in career-level fire suppression. They are offered here to all those in the fire service, career and volunteer alike, whose job it is to manage the transition to a gender-integrated workforce and who wish to make that transition as smooth as possible. This book is particularly created for, and dedicated to, fire service managers. The ease of that transition depends above all on their commitment and hard work.

Many Faces, One Purpose was prepared under contract to the Federal Emergency Management Agency (FEMA) by Women in the Fire Service, Inc. It was made possible only with the assistance of dozens of people from fire departments and other agencies throughout the country who provided information and shared their valuable insights. For this assistance, the researchers and writers of this handbook offer their sincere thanks.

Women firefighters: a status report

Nearly 25 years after women first entered firefighting as a career, more than 4,500 women hold career-level fire suppression positions in nearly a thousand fire departments in the United States. Hundreds more work for the Federal government or State agencies in wildland fire suppression roles. Overseas, women are career firefighters in Canada, Great Britain, France, Germany, the Netherlands, Denmark, Australia, New Zealand, South Africa, Costa Rica, Panama, and Brazil. Many thousands of other women, in the U.S. and elsewhere, work in the fire service in nonsuppression roles: emergency medical services (EMS), fire prevention, inspection, arson investigation, communications, and public education.

More than 900 U.S. fire departments employ women firefighters and very few major departments have yet to hire their first woman. While nationwide only about 2 percent of firefighters are women, many career departments' percentages are three or four times that, and a few departments' ranks are 10 to 15 percent female.

Women's history as volunteer firefighters is much longer than in the career sector, reaching back well over a hundred years. While reliable numbers are difficult to obtain, it is estimated that among the volunteer and paid-on-call fire and rescue forces in the U.S. are perhaps 40,000 women firefighters, and still more women who serve as emergency medical technicians (EMT's) and paramedics.

Women are found in all ranks of the fire service, from recruit firefighter up to chief of department. Women fire chiefs lead organizations ranging in size from small volunteer departments up to those that protect cities the size of Madison, Wisconsin; county departments such as Cobb County, Georgia; and comparable agencies within the wildland fire service. The first generation of career women firefighters, which entered the fire service in the mid- to late-1970's, is coming of age, and the number of career fire service women at the chief officer level increases every year.

There is no such thing as a "typical" woman firefighter. Women firefighters come from all backgrounds, races and ethnicities. They may be single, partnered, married, divorced, or widowed. They may be 6'2" and weigh 200 pounds, or 5'1" and weigh 110 pounds. They may have no children, or be mothers or grandmothers. They may be as young as 18 or as old as 70. They may have a high-school education or Ph.D. What this diverse array of women firefighters has in common is their dedication to their work and their commitment to serving their communities through the fire service.

Contents

Introduction i

Women firefighters: a status report iii

Recruiting women firefighters 1

Recruiting women as volunteer firefighters 17

Physical performance testing for firefighters 19

Legal aspects of physical performance testing 31

Firefighter training 35

Stopping sexual harassment in the fire service 41

Sexual harassment: the legal background 55

Reproductive safety and family issues for firefighters 65

Child care for the fire service 75

Nepotism and firefighter marriages 77

Fire station facilities 79

Hair and grooming standards for firefighters 83

Promotional issues for fire service women 87

Supporting workforce diversity 91

Firefighter protective gear 109

Manufacturers and distributors of firefighter protective clothing 118

Resources 123

Appendix: legal issues 125

Bibliography 131

Recruiting women firefighters

When fire chiefs talk about firefighter recruitment, they usually mean the effort that is spent, shortly before an application period opens, to get candidates to apply for job openings. Recruitment is the way a fire department attracts new members. Fire chiefs who wish to diversify their department's workforce or attract specific groups of people (college graduates, licensed paramedics) have learned to target those groups in the recruiting effort. This section of the handbook outlines the basic elements of targeted recruitment programs and presents some insights from programs already in place.

A productive recruitment drive is just part of what it takes to increase the number of women on a fire department. For recruitment to be really effective, managers must establish a positive climate within the department before encouraging women to become firefighters. Fire departments also must begin to recognize and take advantage of the recruitment impact of most of their public activities. Expanding the concept of recruitment in these two directions will make the recruitment drive itself more productive and will increase the likelihood that the women who are recruited actually will become firefighters.

Before recruitment begins

The skills and dedication of the people working in the recruitment unit, the creativity that goes into designing the program, and the verbal, logistical, and financial backing given to the effort by top management all play important parts in the success of your department's recruitment drive. All of this effort and investment must be supported. Your recruiters' message will be that your fire department wants women to become firefighters. But if other aspects of your department give out a conflicting message, or if the department is unprepared for a workforce that includes men and women, much of your recruitment effort will go for nothing.

A recruitment effort has not necessarily succeeded just because dozens or even hundreds of women fill out applications and show up to take the test. All too often, recruiters and administrators focus only on sheer numbers of women candidates. Giving greater consideration to what you're really trying to accomplish will show the benefits of a longer-range view. The true measure of the success of a recruitment drive aimed at women is found much farther down the road, in the number of women who are on the job as skilled and productive firefighters 2 or 3 years later. This has very little to do with how many women originally showed up to take the test.

Hiring women who will become good firefighters and stay on the job involves efforts outside the recruiter's realm. The out-front recruitment effort is just the tip of the iceberg. Its long-range success rests on work that must be done elsewhere in the department. If your recruiters are to spend their time making it known that your department wants women to work there as firefighters, you as chief must make sure that's really true before they ever begin.

A prerecruitment checklist

This checklist covers some of the preparatory steps that fire department managers can take to make the recruitment of women firefighters more effective, and to help ensure the retention of women who join the department.

The application and testing processes

- Has the testing or other selection process been established in full detail?

- Will candidates be informed about the components of the process, the contents of the physical test, how the tests will be administered, how each component will be scored, what constitutes a passing score, and how the final ranking on the hiring list will be determined?

- Has the entry-level physical test (if used) been validated for your department, based on a job analysis conducted by an outside expert? Does the scoring system reflect the test's accuracy in predicting job performance? Has the pass/fail point been set by testing a random sample of firefighters already on the job? Have skill-dependent items on the test been eliminated or minimized?

- If test practice sessions will be held, are personnel who will staff the sessions familiar with techniques that may be more effective for smaller or shorter candidates? Will a variety of safe and effective techniques for performing the test events be demonstrated to candidates and permitted on the test? Have personnel who will administer the test been instructed as to which techniques will be permitted?

- If candidates will wear protective gear or other specialized items during the test, are these available in sizes to fit all candidates?

- For fire departments that do not give an entry-level test, has the application process, including both its formal and informal elements, been reviewed to make sure it doesn't tend to screen out women?

[See section on physical performance testing, page 19.]

Policy development and review

- Have an anti-harassment policy and confidential complaint procedure been put in place? Has training been held to educate all personnel about harassment and the department's policy?

- Are policies in place to protect the reproductive safety of all personnel with hazardous duties? Have employees received information about the reproductive risks of firefighting?

- Have departmental policies affecting the employment of relatives been reviewed for their impact on married couples?

- Have departmental policies on hair length and personal grooming been revised as needed?

[See sections on sexual harassment (page 41), reproductive safety (page 65), firefighter marriages (page 77) and hair/grooming standards (page 83).]

Recruit training

- Does the training staff have a positive approach towards its job, and towards training women? Is the staff knowledgeable about cultural barriers that may exist for women on the job?

- Have interested women firefighters or officers been assigned to the training academy as instructors or assistants?

- Have the training curriculum and methodology been reviewed for their applicability to a diverse group of students? Are instructors prepared to instruct students with a range of learning styles?

- Will recruits be given a clear understanding of performance criteria? Will skills be taught and recruits be given ample opportunity to practice them before being evaluated?

- Does the training staff have a clear procedure for evaluating and documenting recruit performance and for dealing constructively with any problems?

[See section on firefighter training, page 35.]

Fire station facilities and firefighter protective gear

- Are uniforms available in cuts and sizes to fit women?

- Will properly fitting protective gear be available to all fire recruits from the time it is first needed in training?

- Do all fire stations have adequate facilities for a workforce that includes both women and men? If not, has a plan or timetable been implemented to bring them up to standard?

[See sections on station facilities (page 79) and protective gear (page 109).]

What is clear from this list is that most of the issues that arise when a fire department becomes gender-integrated should be considered long before a targeted recruitment of women ever begins. A fire department that has little management commitment to diversity, an unvalidated, speed-to-completion entry-level physical test, hair-length standards based on men's fashions, or a "We'll deal with it when it happens" approach to firefighter pregnancy, shows it truly does not care what happens to any women who might join the department.

People are unlikely to be motivated to enter a workplace in which they are unwelcome. One fire chief put it succinctly. "It was hard to get women to apply for the job, because the men didn't want them there." Women firefighters already on the job may be reluctant to join in a recruitment effort and may even actively discourage other women from applying for the job, since they know the problems women in the department face. It is both practically and ethically difficult to recruit women into such an environment. It is a waste of effort, or worse, to recruit women before the above issues are addressed.

Recruitment: the big picture

A fire department's real recruiting effort, like its public education work, goes on all year round. Volunteer fire departments, many of which are in constant need of new members, already know this. Consciously or otherwise, a fire department recruits new members and gives out information about itself all the time. A chief who takes advantage of the recruitment potential of everything the department does can make its efforts both efficient and productive.

Women and people of color currently on the department should be included in all of the department's public activities. Firefighters self-recruit--attract more people like themselves--because of their visibility. If only white men are visible when your department puts out a fire, holds a rescue demonstration, or has a press conference, it's primarily white men who will be recruited as a result. Every time your department is in the public eye, and especially when it is being covered by the press, its diversity should be visible.

Maintain a consistent commitment to nonsexist attitudes and language. Using and encouraging gender-neutral language by department members and others is an important starting point. If your local newspaper still refers to firefighters as "firemen," a letter to the editor from the fire chief not only will help get the practice stopped but also will demonstrate your support for women on the job.

In addition to maximizing the department's public appearances for the benefit of recruitment, specific steps (described below) can be taken to encourage potential candidates to consider and prepare for firefighter jobs. These long-term efforts will produce results over the course of years. Short-term efforts to recruit women can be difficult because many women simply have never considered becoming firefighters; it's a hard decision to make in a short time. The more information about the job you can provide to possible future employees far in advance, the easier it will be to locate and attract good, qualified personnel when you need them.

Vocational counselors

Develop and maintain a good working relationship with high-school guidance counselors and with career-placement personnel and vocational counselors at the colleges and universities in your area. These people can be where you cannot: in touch with young people who are making career decisions. Are the counselors in your high schools and colleges encouraging young women to consider fire service careers? Make sure they are aware of your department's interest in hiring women firefighters, and provide supplies of literature, videotapes, and other information for them. Invite them to orientation sessions, or hold special sessions for counselors to discuss the ways you and they can work together.

Introductory programs

Explorer posts, job shadow programs, high-school "cadet" programs, ride-alongs for community members, and resident student programs all can be productive ways of recruiting new firefighters who, by the time they complete the program, will be familiar with your department's operations and personnel, and may possess basic firefighter training as well. These are excellent recruitment tools that can be effective at recruiting women if you make sure women are fully included in them and encouraged to participate.

Extended contact with potential candidates

Many traditional elements of a short-term recruitment drive can be kept operating easily year-round. Periodic open houses, practice test sessions, and orientation sessions can draw candidates' interest well ahead of test dates, allowing more time for them to develop their strengths and skills or to seek relevant education.

The fire department or personnel department should accept job interest cards at all times, even when a test is not planned. To keep the database current, these can expire in a year or two, at which time a card is sent out to verify the individual's continued interest. A verification card also is sent out if a hiring process opens up and the person does not apply; those who do apply would be kept on the list automatically for another year or two, unless they are hired.

The recruitment program

Some fire departments do a generalized publicity effort when their test is announced or, in the case of volunteer departments, when more firefighters are needed. The purpose of the publicity is to get job information to as many potential applicants as possible. Many departments, however, have found that large numbers of white men will apply for firefighter positions even if the openings are not publicized. This can be because of a strong family-and-friends tradition within the department or, more generally, because white men as a group are well aware the fire service is a career option for them.

A department that wishes to increase the diversity of its workforce may decide instead to focus its limited resources of time and money on a recruitment effort that targets groups under-represented in the workforce or applicant pool. While no one is discouraged from applying for the job and no applications are rejected, the natural tendency for the existing workforce to self-recruit is counterbalanced by publicity aimed at those who might not otherwise apply. Without deliberate intervention in the form of targeted recruitment, the majority of applicants will continue to be from the existing majority group in the fire service, white men.

Setting up a recruitment program involves planning, commitment, creativity, and often the coordinated work of a number of people from different city departments. A recruiting drive can be as basic as one person with a slide show working for 3 weeks, or as complex as a fully staffed division operating over the course of a year. The exact details of the project will vary considerably depending on several factors, including

- the size and resources of the particular fire department;
- the goals of the recruitment program;
- the amount of time available for recruiting;
- the make-up of the community and the labor pool; and
- the skills and commitment of the people who design and deliver the program.

Whatever its dimensions, a well-designed and carefully organized recruitment program always will achieve greater success than one that is haphazard or based on misconceptions. Certain key elements will be present, in one form or another, in any effective and successful recruitment drive. These elements, discussed in detail on the following pages, include

- management support;
- careful recruitment team selection;
- realistic schedule;
- targeted recruitment materials;
- publicity within the community;
- effective use of the media;
- orientation sessions and open houses; and
- physical test preparation sessions.

Management support

Fire department leadership must give its firm support to the recruitment and integration of women into the department. All aspects of the recruiting effort must reflect management's sincere commitment, not only to bring women firefighters onto the department, but also to support a diverse fire service workforce. Management can support recruitment in the following crucial ways:

- Obtaining the funding necessary to make the program a success--by making it a priority within the department's budget, by seeking additional city funds (such as from the personnel department or other agencies concerned with equal employment opportunity), or by obtaining donations from sources in the community. Free technical assistance from community organizations funded by State or Federal job training programs also may be available.

- Making other resources available in order to maximize the allocated funds. This might include vehicles, office space and equipment, or the reassignment of fire personnel and support staff. Resources also can be borrowed from other city departments, which may be able to provide prior applicant lists for other nontraditional jobs, photographic and darkroom work, audiovisual equipment and expertise, and so forth.

- Working with other city departments, if necessary, to get adequate lead time (the time between the day the test is announced and the day it is given, during which most recruitment will be done). Many firefighter recruitment efforts are doomed from the start because the fire department cannot get more than four to six weeks' notice before a test will be given. The fire service manager also should help settle any jurisdictional or procedural problems that affect recruitment.

- Demonstrating leadership by representing the program positively to elected officials in order to obtain their support, and by making public statements, particularly in the media, in support of the recruitment effort and of hiring women and people of color.

Public statements should not emphasize numbers as a measure of the success of a recruitment effort. If potential candidates and incumbent firefighters perceive (correctly or incorrectly) that management just wants to hire women to get numbers to fill a hiring goal, the sincerity and effectiveness of recruitment will be undermined severely. Saying "We want to hire 10 women" implies two things, and both of them are negative: first, that you will hire 10 women just to hire women, even if not all of them are qualified; and second, that if more than 10 qualified women apply, you will not hire the others. It also can make your recruitment drive appear to have been a failure if you only end up hiring nine women.

Instead, make a positive "goals statement" that emphasizes your commitment to diversify your firefighting workforce and to support its diversity in meaningful ways. For example, one fire department said in its job announcement:

We are looking for professionals who want to be part of a progressive, innovative fire department. Our goal is to have a workforce that reflects the diversity of our community. Women and people of color are especially encouraged to apply.

Recruitment team selection

Select members of your recruitment team who have qualities that will make them effective. Don't choose recruiters for reasons that make sense from a limited perspective but are irrelevant or unproductive when it comes to recruitment, such as:

- They always have been in charge of recruitment.

- They are injured or pregnant and need a light-duty assignment.

- They belong to one of the groups being targeted for recruitment.

All of these practices can create problems. People who have "always" done recruitment probably will continue to produce the kind of candidates they've produced in the past. If these are not the people you're looking for, it could be time to make a staffing change. A firefighter who has been removed from the line for injury or some other reason is usually chosen for the department's convenience, not because of his or her qualifications to be a good recruiter. (If such people should happen to possess the skills you need, however, they certainly should not be overlooked.) And although it is less obvious, the same is true of firefighters or officers who belong to the targeted groups.

Women firefighters who have an interest in recruitment should be used at orientation sessions and other public-contact points where women who are potential candidates will want to hear from, and ask questions of, women on the job. But a good speaker or advocate is not necessarily a good program manager. Women firefighters assigned to coordinate recruiting programs often have few credentials for the job and, in some cases, little interest in it. Use the valuable abilities and enthusiasm of women who want to be involved, if they have the abilities you need, but don't assume someone will be a good recruiter just because she's a woman and a firefighter.

Identify the skills you will need on your recruiting team before you (or the person in charge of recruitment) pick its members. Interest in, and commitment to, the recruiting drive are prerequisites: no one should be chosen who does not want to be involved. Useful skills and traits include

- education and experience in marketing and public relations;
- organizational skills;
- graphic arts skills;
- writing skills;
- public speaking skills on camera or radio, and before groups;
- computer literacy;
- bilingual ability, if your community is ethnically diverse;
- ties with targeted groups; and
- reliability and the ability to meet deadlines.

No one will possess all of these traits. Diversity on the team is important for that reason as well as to provide the flexibility and the range of creativity that will permit a variety of approaches. The recruitment coordinator and other top recruiters also must have good organizational skills and be able to work well together in a concerted effort.

The people involved in the recruitment program may come from various areas. They may include firefighters and nonsuppression personnel, support staff from the fire department or other city or county departments, community volunteers, and members of other fire departments (if your own department has few or no women, or if you are participating in joint recruitment). Smaller fire departments also may need to borrow the clerical or computer services of other city departments.

Whatever the size of your department, tap all of its resources. A recruitment program is very similar to a public education effort, and your public education division should be a gold mine of assistance. Its staff knows how to scale a motivational message to a target audience and present that message in a way the audience will understand and respond to. Similarly, your public information officer has skills and contacts that can be very useful in designing and distributing press releases and in getting media coverage of recruitment events.

Use community volunteers to distribute literature and to make contacts with various community groups. Small departments with limited budgets also may be able to find people who will donate their professional skills to design a brochure. Local businesses might donate all or part of the cost of printing literature and posters for the recruitment drive. Cable-access television channels may provide video production equipment and editing facilities. Fitness centers and gyms may be willing to offer discount memberships to firefighter applicants preparing for the test.

Setting up the schedule

The recruitment coordinator is responsible for the program's overall design. This means determining what types of fire department and community resources are to be used, what kind of media publicity will be required, what the priority markets are, and what is to be done when. The first step is to identify all of the tasks to be completed. At a minimum, the recruitment team will need to

- Develop the budget for the recruitment program, based on total allotted funds and any donated resources.

- Identify useful community resources, make preliminary contact with key people in each group, and get the dates of their regular meetings and any pertinent special events.

- Identify career fairs and similar events scheduled during the recruitment period.

- Write brochures, test information, and other literature, and have them printed.

- Design and print posters and other items requiring graphic art or photography.

- Write television and radio public service announcements (PSA's), newspaper advertisements and press releases; arrange for newspaper articles and television news coverage.

- Announce the recruitment drive through any online connections your department has, and on your city's Web page, if it has one.

- Produce videotapes for orientation sessions and test familiarization.

- Schedule orientation sessions.

- Set up "open house" dates at fire stations.

- Set up physical performance test practice sessions and written test study sessions.

- Get the necessary logistical support: vehicles, office space and supplies, phone lines, voice mail, fax machine, chairs, audiovisual equipment, security for office use at night, etc.

Arrange the tasks into a schedule, timeline, or action plan. This should include when each task must be done, who will do it, how long it will take, and what resources will be needed. Allow time for unforeseen delays; don't schedule a major event for the day after needed materials are due from the printer. If bad weather might cancel an event, schedule an alternate date in advance.

Recruitment efforts must be given adequate time in order to be effective, particularly when the recruitment is aimed at women. Deciding to become a firefighter is not an easy decision for many women. Issues of self-image and self-confidence, a partner or spouse who may not be supportive or understanding, questions about child care and having to be away from one's family for 24 hours at a time, and the risk involved in leaving a job to start something new and uncertain all frequently arise and cannot be resolved quickly. Women also may need time in which to prepare for the physical test and the physical demands of the job.

If you are limited by decisions made by others--for example, Personnel or Civil Service sets the test date and will give you only a few weeks' notice--many of the above items can be developed in advance and kept on hold until the date is known. Write the program budget, contact community resource people, select the recruitment team and coordinator, establish a timeline, and draft all written materials, leaving the date and other undetermined factors blank. If the physical test will not be changed before it is given, make the videotape that explains and demonstrates it. Contact reporters from the local papers and television stations to let them know of the events they will be able to cover once recruitment is underway. Fire departments that cannot obtain adequate lead time for an effective recruitment drive also must rely heavily on the ongoing, year-round types of recruitment discussed earlier.

Recruitment materials aimed at the target group

In designing and writing your recruitment materials, remember you are trying to appeal to a different group of candidates from those who have traditionally applied for the job. Recruiting women to become firefighters is not just a matter of going to places where there are lots of women and handing out the usual information about application deadlines and test dates. The content and image of your message must be different and must address cultural preconceptions that women have about themselves and the job. A brochure or insert specifically aimed at women candidates can be highly effective.

Your primary piece of recruitment literature should be a brochure that presents information about three things: the job of firefighting, your fire department, and the upcoming testing process. The design should be simple enough that you can afford to have hundreds or thousands of copies printed up for wide distribution.

Some departments have their literature produced professionally and find the results well worth the investment. If no funds are available for this, it is possible to put out a high-quality product in-house. A compromise is to have a single sheet or folded flyer produced professionally that contains information that will not change (about the job and the department), and add to this the material produced by the department. Because the cover sheet or exterior of the flyer can be used for several years, you will not have to reinvest in its production, and you can take advantage of volume discounts in the original printing. Your own added material, which contains information about the next test, current salary and benefits, etc., keeps the literature up to date. Leave room for an address and postage on the outside of the brochure to make it easy to mail.

The brochure should explain all aspects of the firefighter's job, including its rewards and its demands. Women are attracted to firefighting for many of the same reasons men are, such as the challenge of a physically demanding job, the rewards of performing a service to the community, and good pay and benefits. All of these should be emphasized.

In writing your brochure, keep in mind how little the average nonfirefighter knows about firefighting and the other work firefighters do. Women in particular may have inaccurate preconceptions about the job that either can keep them from applying or lead to problems later on, when the job turns out to be different from what they expected. Do not understate the risks involved in the job, but don't overemphasize them, either. Make it clear that all recruits will be fully trained before ever having to deal with an emergency, and that safety is always a priority. Discuss recruit training: how long it lasts, what the schedule is, what is taught. If your station facilities are designed to accommodate a workforce of women and men, be sure to mention that.

Use gender-neutral language in the recruitment brochure. This means not referring to firefighters as "he," substituting terms like "staffing" for "manning," and saying "women" and not "females." Seemingly little things contribute in big ways to the impression you make. The overall message should not be that women can manage to perform the job of firefighter and somehow fit into a "man's" job, but that women are a valuable asset to the fire service, and can enjoy productive careers there. Photos or drawings in the brochure should include people of both sexes and a variety of ages, sizes, and ethnic backgrounds.

Your literature should include the customary information about pay, hours and benefits, pension plan, number of stations the department has, average call load, and so forth. It also should provide clear explanations of the following items:

- details of the application, testing, and hiring processes, including age limits for hiring, eyesight requirements, medical examination, and drug screening;
- how and where to get application forms;
- whether applications will be available by mail or can be picked up by one person for another;
- what to bring if applications must be picked up in person (for example, proof of identity, citizenship, or city residency);
- whether picking up or turning in an application early is important, such as in the case of a scoring tie or if the number of applications to be given out is limited; and
- whether applications must be filled out on the spot, or may be returned by mail.

Information about the testing process should indicate which test will be given first, and when and how applicants will be notified whether they have passed and if they are to go on to the next step in the process. If handbooks or other study materials are to be distributed for the written test, include information on how to obtain them.

Explain in detail what the physical performance test entails and how it will be administered and scored. List the dates and places of test practice sessions, or provide a phone number to call for this information. If child care will be available at practice sessions or at the test itself, mention that. Provide information on other resources available through the fire department or elsewhere, such as through the union, other firefighters' groups, and the community. These might include

- a videotape about the physical performance test that explains all of its components and demonstrates techniques that may be used to accomplish them;
- weight-training classes or gym memberships;
- current firefighters who will serve as mentors to candidates;
- open-house dates, or other opportunities for station visits; and
- test-taking study sessions, support groups, or assistance in finding a fitness training partner.

All recruitment literature, whether contracted out to a private company or produced by one firefighter with a desktop publishing program on a home computer, must be neat and professional. Literature that looks sloppy, contains misspellings and grammatical or typographical errors, and generally appears to have been done hastily and given low priority, makes a negative statement about your department and your commitment to recruiting women. Attractive, professional materials can be produced at relatively little expense.

Have someone who is not in the fire service proofread the material, not only for errors but to see if it makes sense to someone unfamiliar with the field. If an outside agency produces your literature, check their text for correct use of terminology. Most nonfirefighters don't know that a fire engine is not a fire truck, and the fire service itself doesn't agree on what a "rescue unit" is.

Posters. Fire departments that have asked applicants how they learned about job openings have often found posters are a low-percentage effort. (The winners were newspapers, direct mailing of literature to potential candidates, and knowing someone on the department.) Posters are usually produced outside the department, due to the technical demands of the process. If you do go to the expense of producing and distributing them, they should be of high quality. The typical poster consists of a color or black-and-white photo and a caption such as "Can you fill these shoes?" (with a photo of fire boots), "We're looking for a few good women," or "It takes all kinds to make a fire department."

Posters may be effective recruiting tools if they are well designed to appeal to your target group, and if they are placed in the right locations. They may work better in smaller towns than in big cities where there is more competition for people's visual attention. They may possibly be more effective for long-range recruitment, if they are posted where they can remain for some time and more people will see them. As a courtesy to the businesses or other agencies where you place the posters, send someone around to collect them once the information is out of date. Posters placed where they are subject to vandalism or graffiti should be checked periodically for replacement or removal.

Videotapes. Videotape technology has created a revolution in information-sharing that the fire service has only begun to exploit. It is relatively inexpensive to produce videotapes and make them available to support your recruitment drive.

One tape should deal with your entry-level physical test, demonstrating each element of the test both separately and as it falls into the testing process. Use women and smaller men among those demonstrating the test items and evolutions, particularly the events that are most affected by height, leverage, and technique. Make sure your test will not change after the videotape is made; the tape must be accurate and give as much helpful information as possible. The tapes should be available on loan. If your department does not wish to handle this, you may be able to negotiate an agreement with local video rental stores.

You also may wish to develop a videotape for use at orientation sessions and other informational events such as booths at career fairs. This tape should provide information about the job, and about being a woman firefighter, from the woman's perspective. It might include footage from actual fires, training sessions, "life in the station" scenes, and interviews with women firefighters on the job and at home. Be sure to include more than one woman, to show a diversity of backgrounds, sizes, ages, and personalities.

Publicity within the community

Following are examples of locations or organizations where it may be most effective to publicize job openings, place posters, leave flyers, and speak to interested groups or to individual women.

- colleges and universities*;
- high schools, if your minimum age is 18;
- athletic clubs, teams, and events: universities, colleges, and community colleges; women's softball, basketball, and volleyball leagues; runs and triathlons; women's powerlifting and body-building competitions; self-defense training schools;
- gyms and fitness centers;
- career fairs;
- military bases and discharge centers;
- factories and union offices;
- women's advocacy groups, women's commissions, YWCA, 4-H, Girl Scouts, American Association of university Women, etc.;
- minority-employment and tradeswomen's networks;
- church and other groups in specific ethnic communities;
- other departments within your city or county government: water, parks and recreation, sanitation, etc.;
- agencies that offer fire recruit training;
- volunteer fire departments and smaller career departments, particularly if your department does not have a residency requirement; and
- wildland firefighting agencies, particularly if you are recruiting towards the end of the fire season (in the fall).

Use your current personnel as recruiters, not only with the public but with people they know personally. Family members and friends of current firefighters and police officers, and nonsuppression employees of the fire department are two key sources of potential recruits. These people are somewhat familiar with the demands and rewards of the job already, and often with how the department functions.

---

*You may wish to target specific academic areas that you feel will benefit your department. Look beyond fire science programs to social services, education, and business or public administration. Fire departments that provide wildland fire protection might look for candidates with forestry backgrounds.

Effective use of the media

The media are the least expensive and most useful resource for your recruitment effort. At very little cost to you, they will deliver your message into the homes of thousands of people. Not only should you use them for classified ads and public service announcements, but often they will be willing to cover your recruitment events as news or feature stories. Small-town newspapers are especially useful, as they are often in need of material, and their subscribers often read the paper in great detail.

Some fire departments have gotten the media to cover their recruitment drives by inviting a woman reporter to go through the physical test, or to spend a day in the station. This can be a gamble: if the reporter can't complete any of the test events, the job will seem off-limits to women. Providing her with some preliminary training on the events, or suggesting that the station or paper send a reporter who is physically fit and active, may help. Even better, have a woman firefighter go through the events (successfully) at the same time the reporter does, especially if cameras are present.

Encourage reporters to promote the idea that being a firefighter is neither easy nor impossible for women; point to the number of skilled and successful women already on the job in your department or in the area. Reporters usually will be grateful for a fact sheet about women firefighters, to use as background for their stories. Providing this information will help prevent the media from portraying women firefighters as though they were unusual, even if the women you're recruiting will be the first ones in your department.

Local cable-access channels usually will show your orientation videotape if you provide them with a copy. They may be willing to show the tape in conjunction with an interview or call-in show where representatives of the fire department discuss firefighting as a career for women and give information about job openings.

The broadcast media are required to dedicate a certain amount of free air time to PSA's. Fire departments can take advantage of this to publicize their recruitment drive. Videotaped television spots should feature visuals of women firefighters on the job. Radio announcements and television spots that do not have visuals should use women's voices. The text of your PSA's should be consistent with your written handout materials and, in fact, much of it can come directly from your literature.

If your department is large, has an attractive location, offers particularly good pay and benefits, or is able to hire firefighters from elsewhere on a lateral-entry basis, consider advertising nationally in fire service trade journals or women firefighters' publications. Keep in mind, though, that monthly periodicals often require a lengthy lead time; be prepared to submit material to them well in advance. Women firefighters' organizations also may be able to provide you with recruitment brochures, videotapes, and other helpful materials.

Orientation sessions and open houses

Speaking to groups of potential job candidates is a key element of recruitment. Hold orientation sessions for potential candidates at times and locations convenient to them. Providing child care during the sessions will not only will make it possible for more women to attend, but also will demonstrate your department's commitment to hiring women and its awareness of women's needs.

Speakers at the sessions should include firefighters and officers, both women and men, of varying ethnic and personal backgrounds. Their material should include a basic description of what firefighters do, in down-to-earth, nontechnical, and unglamorized terms. It also should describe the testing and training processes. Women firefighters should talk about their experience on the job: what it's been like for them and why they enjoy the work. A high-ranking officer should be present to reaffirm the fire department's commitment to cultural diversity and equal employment opportunity. All speakers should be positive about the job, honest about its demands and accessible to candidates' questions. Showing your recruitment videotape at the beginning should reduce the amount of time needed for questions. You also may wish to have firefighters' protective equipment or fire apparatus on hand for candidates to see and ask questions about.

Distribute applications to all who are interested, if your personnel system allows you to do so. If not, it may allow you to distribute interest cards by which individuals can request that an application be mailed to them. If even this is not possible, have each attendee fill out a card with his or her name, address, and phone number. The cards should be coded so you can track how productive each event was.

Open-house sessions provide a chance for candidates to come into selected fire stations, view the apparatus and facilities, and talk with firefighters and officers. When scheduling these events, consider the suitability of the particular crew, the convenience of the time and location for the target audience, the type of apparatus and station responsibilities (self-contained underwater breathing apparatus (SCUBA) team, haz mat unit), etc. Recruitment personnel should be present as well, particularly if you don't plan to take the station out of service. Some fire departments offer open houses on a regular basis as part of their year-round recruitment effort, as well as for community and neighborhood relations.

All of your fire stations should function as recruitment outposts throughout the recruitment drive. Copies of recruitment literature and, if possible, applications, should be available at all stations. All department personnel should be able to answer basic questions about the testing and hiring processes.

Physical test preparation sessions

In some fire departments, the testing procedure alone takes many weeks. Months or even years can elapse between the first orientation session and the day a new firefighter actually is hired. Fire departments unnecessarily lose many good candidates during this time. Throughout the application and hiring process, the more contact you can maintain between women candidates and the fire department, or among the women candidates themselves, the less likely you are to lose them, and the more likely they are to maintain their interest and motivation. Test preparation sessions for applicants are one way to maintain contact as well as to help candidates prepare for the upcoming testing process. They also accomplish several other important goals:

- They allow candidates to measure their fitness levels against what the fire department will require of them, and to identify any areas of weakness.

- They provide instruction for candidates in the techniques that can be used most effectively to accomplish the test items.

- They give candidates experience in handling the equipment used on the test.

- They give the fire department an opportunity to identify any problems or inconsistencies in the design and administration of the test.

Some fire departments offer both their physical test and practice sessions on a regular basis: for example, practice sessions monthly and the test twice a year. Most departments, however, test only every year or two, and hold practice sessions only during the weeks or months right before the test. In either case, the practice sessions should be held far enough in advance that candidates can go through a practice to find out their areas of difficulty, and still have time afterwards to improve in those areas before they actually take the test.

In setting up practice sessions, consider the following:

- The equipment, tasks, and sequence of events should be the same as on the actual test, or as close as possible. It may be convenient to set up just part of the test for each session, but if your test consists of a timed sequence of tasks, the only way that participants can get a realistic idea of the endurance needed to complete the test in the allotted time is if they can go through all of its events.

- Obtain release forms from all participants prior to any hands-on activity. You also may require that certain types of clothing or footwear be worn, or you may wish to provide helmets, gloves, or other protective items for candidates to wear during practice. If you are providing the equipment, make a full range of sizes available to ensure a safe fit for all participants.

- Firefighters and officers who will staff the practice sessions should be selected and trained carefully. They should support and encourage all candidates, and should offer instruction in all techniques that will be acceptable on the test. If candidates attempt to use techniques or methods that are clearly unsafe, the staff should inform them that their method is not acceptable. It is very important to have consistency among the personnel giving instruction at the practice sessions and those who will administer the test, so candidates receive reliable information.

- Schedule the sessions for different times of day and different days of the week, to maximize the number of candidates who can attend. If the testing equipment is relatively portable and easy to set up, consider holding practice sessions at several locations, for the same reason.

Physical test preparation: other resources

Drawing on the resources of a local university, some fire departments have had exercise scientists create a task-specific training program designed around the items on the physical test, the physical demands of firefighter training, and the physical requirements of the job. (If assessing those three items--the test, the training, and the job--produces three different sets of physical requirements, it is probably a good idea to re-evaluate your testing and training processes.) Describe the program in a way that will be clear to someone unfamiliar with weight-training terminology.

Look for community support to supplement what your department can do. Gyms and fitness clubs may be willing to offer membership discounts to candidates who are preparing to take the firefighter exam; make sure the staff members of these gyms receive copies of your training program. Donations from businesses or small grants from community funds may be available to sponsor applicants' gym memberships or tuition for a fitness-training course at a local college.

Other forms of preparation

Many fire departments, union locals, and support groups offer study-skills and test-taking workshops that help candidates prepare for written entry-level tests and for the fire academy. These classes or sessions focus on reading retention, problem-solving, and other skills and tips that help candidates in the testing process. Community colleges often have useful resource people in these areas: you may be able to borrow a faculty member to conduct workshops, or find ways to fund candidates' tuition in a study-skills course. This can be of great benefit to women who are re-entering the job market and to those with limited academic backgrounds, particularly if the written test is especially competitive. Workshops on interview skills also can be helpful in preparing candidates for that portion of the hiring process.

A mentoring program can be set up to provide direct support by incumbent personnel of firefighter candidates or recruits. Mentors offer support, a point of contact within the department, technique tips, and additional needed information. Individual firefighters or officers, through the department or the union, may volunteer to serve as mentors to candidates, trainees, or probationary firefighters. A list of names should be offered to the candidates or recruits so that they can choose the firefighter they personally are most comfortable contacting.

Program evaluation

When the recruitment effort is over, evaluate the program's effectiveness. The applicants themselves can provide direct and constructive feedback. A space on the application form asking how the person heard about the job, and an evaluation form to be filled out by attendees at orientation sessions or viewers of videotapes, can help you determine where your efforts can be improved.

Save the database you have developed on potential candidates and use it as part of your mailing list for the next recruiting drive. If your eligibility list is likely to last for some time, implement a way to maintain contact with women who are on the list but haven't yet been hired. This will let them know they haven't been forgotten, and will increase the chances they'll stay interested in the job.

Each member of the recruitment team, and other people who had a significant part in the effort, should write a report. This should be followed by a meeting of everyone involved. A final report with recommendations for change and an action plan for implementation of the changes should be forwarded to the fire chief or appropriate member of top management. Followup letters of thanks from the recruitment coordinator or fire chief to all community members who donated services or money are not only polite but make future donations more likely.

Recruiting women as volunteer firefighters

Women's participation in volunteer fire departments in the United States goes back nearly two centuries. Women are known to have been volunteer firefighters in the early 1800's, and whole volunteer companies of women were enlisted to provide fire protection on the outskirts of cities such as Los Angeles in the 1920's.

Volunteer fire departments vary widely in their attitude toward women firefighters. At their best, they welcome anyone who has the abilities and dedication required to be a firefighter. Many women receive strong support from the men on their volunteer fire departments, and are able to use their talents to become firefighters, paramedics, fire officers, and chiefs.

At the other end of the spectrum are volunteer fire departments that adhere rigidly to a "boys' club" tradition, excluding others either by specific prohibition or by the force of custom. People of color and women who do manage to join departments of this type often face multiple barriers: isolation, denial of training, refusal to provide properly fitting protective gear, and more overt forms of harassment.

Volunteer firefighters outnumber career firefighters in the U.S. by a ratio of three to one. Thousands of communities, and thousands of square miles of rural areas, are protected by volunteer fire departments. Nonetheless, the number of volunteer firefighters is shrinking. Statistics from the National Fire Protection Association (NFPA) showed there were 10 percent fewer volunteers in the U.S. in 1992 than in 1983. For many volunteer fire departments that are finding it progressively harder to attract members who will devote the required hours to training and emergency responses, the question is no longer "How can we keep women out?" but "How can we get women to join?"

Volunteer fire departments that wish to recruit women and support their involvement in the department should implement and enforce policies against sexual harassment and other forms of discrimination. This is important because the Federal law that protects workers against discrimination does not always apply to volunteers. Some States, such as New Jersey, consider volunteer firefighters to be employees for the purposes of discrimination law; other States do not. [See Appendix for more information.] Therefore, women and people of color must rely on local ordinances and departmental policies to uphold their right to fair treatment in the volunteer "workplace."

Most of the issues discussed in this section of the handbook apply in some form to volunteer departments. The work and traditions of firefighting are the same, even if some of the details and the overall scope of a recruitment effort differ. For example:

- While volunteer fire departments often do not use entry-level physical or written tests, they almost always have some type of selection process that decides who will join the department. Often, this consists of having all members vote on new applicants, a subjective process that can give more weight to the beliefs and prejudices of the existing membership than to the actual qualifications of the applicant. Instead, consider implementing selection procedures that open the department to the widest possible range of qualified people.

- Recruitment in volunteer fire departments often consists only of informal, word-of-mouth efforts, so friends and relatives of current firefighters are those most likely to hear about openings or to be encouraged to apply. Wives, girlfriends, sisters, and daughters of firefighters may be recruited in this way, but for the most part these informal systems only succeed in replicating the existing profile of the department. A more formal recruitment program that makes an effort to reach out to all areas of the community will be more likely to attract a wide range of qualified people.

- Short-term, short-notice child care during emergency calls can be an important issue for women volunteer firefighters, and even more so for couples who are both firefighters. Creative solutions should address this need, increasing the availability of personnel for emergency calls and at the same time demonstrating the department's support for women's involvement.

- Volunteer fire departments usually need to rely heavily on donated resources in putting together a recruitment effort. Fortunately, people in smaller communities served by volunteer departments are often much more willing to donate their time and expertise as a community service than they would to a large municipal fire department.

- Purchasing protective clothing for firefighters is a significant expense for small volunteer fire departments. Often, new firefighters are expected to function in hand-me-down turnout gear used by former department members. This gear often does not fit women. Bringing women onto the department for the first time, or increasing the numbers of women, therefore, can mean a substantial outlay of money for new gear. Some women firefighters have suggested developing regional networks or "banks" for sharing protective gear among departments. Such networks would increase the size of the pool of reserve gear, making it more likely that a pair of size 5 boots, or a coat to fit a 5'3" firefighter, will be available.

Most volunteer fire departments are constantly seeking new members. Giving some attention to the above issues, and adopting the suggestions presented throughout this section to your department's needs, should improve the chances that your department will be able to attract and retain more women firefighters in the future.

Physical performance testing for firefighters

When the first woman to become a career firefighter applied for the job in 1973, the fire department required her to pass a physical test even through they had never before tested firefighters or applicants on their physical performance. The fire service nationwide soon followed suit, as more women began to enter career firefighting positions. As a group, women generally were considered incapable of performing the physical tasks of firefighting. The idea of women firefighters focused attention on the physical demands of the job, and on the need to assess candidate fitness in some way. Many departments toughened their entry-level standards or developed physical tests for the first time. Until a series of court challenges beginning in the 1970's, such tests often were designed haphazardly and administered by untrained individuals with no expertise in testing.

In the 1980's and 90's, fire departments began to address the fitness and performance levels of firefighters already on the job. The impetus for this change came from several sources: the increasing professionalism of the fire service, a heightened awareness of potential liability for injuries and poor performance, a concern for firefighter wellness, the presence of ever-greater numbers of women in the ranks, and the need for an incumbent benchmark to justify entry-level standards. Particularly in response to the introduction of NFPA 1500, *Standard on Fire Department Occupational Safety and Health* in 1987, fire departments implemented on-the-job fitness programs, weight or body-fat standards, medical screenings, and--drawing the most attention--annual performance assessments. In some cases these programs have been voluntary, sometimes with financial or other incentives for participation; often, they are mandatory.

Entry-level physical performance assessments

If there is one area where the fire service is looking for a magic pill, it is entry-level physical testing. The issue is complex, thorny, and often controversial; it raises questions that have no simple answers. There is no perfect solution, no one "best" physical test for firefighter applicants. It is likely that no physical test within the limitations of fire department budgets could predict job performance with pinpoint accuracy.

A good physical performance test will measure applicants against a standard as objectively as possible, without favoring or disadvantaging anyone for reasons unrelated to their suitability for the job. A test that doesn't predict job performance can't select the best recruits for a fire department, and will deny jobs to qualified people.

Most of the time, a fire department will be unaware of the shortcomings of its test. One department, however, learned of them accidentally. It developed a program that allowed a group of women and minority candidates who had passed the physical test, but not scored high enough to be hired, to enter recruit training on a conditional basis, side by side with new recruits hired from the top of the list. In training and afterwards, the department found no consistent differences in performance or ability between those hired from the top of the list and those from near the bottom. The test had completely failed to predict who would be "best" after training.

Designing fitness and performance assessments and validating selection processes, are the province of exercise scientists and other experts. The purpose of this section of the handbook is to clarify some of the legal, practical, and ethical issues affecting physical test design and administration, and provide guidance to fire service managers in assessing the value and impact of a physical performance test.

Types of tests

Approaches to entry-level physical screening for the fire service fall into four general categories: proxy tests; task-based, job-simulation tests; fitness assessments; and no entry-level physical test. Some fire departments' tests combine elements of the first three approaches. All four options offer advantages and disadvantages in terms of their effectiveness and their impact on women candidates. These are discussed below.

Proxy tests. Tests of this type rely on substitute tasks to predict job performance: distance runs to measure stamina, devices to measure grip strength, balance beams to measure equilibrium, and 8-foot walls to measure the ability to scale obstacles. The advantages to proxy tests are that they minimize tasks at which one improves substantially with a small amount of practice, and do not give an inadvertent advantage to candidates who have prior familiarity with firefighting equipment.

When substitute or proxy tests have been challenged, however, fire departments often have not been able to show that the tasks actually measure what they were intended to measure, or have not been able to justify the scoring systems used to evaluate the tasks. For instance, some tests used a dynamometer to measure grip strength. Because the device did not adjust for hand size, it did not accurately measure the grip strength of applicants with smaller hands. When the scoring system for a distance run was examined, in some cases employers could not explain how they determined what would be an acceptable passing score except by caprice or guess.[1] As a result, these tests designed to measure abilities required by the job have been viewed with skepticism by many fact-finders and courts.*

Glossary

Adverse (disparate) impact: refers to a test or other selection device that negatively affects members of a particular applicant group, as compared with members of other groups

Incumbent firefighters: those already on the job (as opposed to firefighter candidates)

Norming: setting the passing score on the physical performance test by administering it to incumbent personnel

Valid test: a selection device (test) from which one can reliably predict job performance

---

Job-simulation or work-performance tests. This is the most common type of test currently used by U.S. fire departments. These tests require candidates to perform simulations of fireground tasks such as carrying tools and equipment, chopping, simulated forcible entry, pulling ceilings, or advancing hose lines, all while wearing protective gear.

Defending these tests as having content validity (*see "Legal Aspects of Physical Performance Testing," page 31*), employers and test developers have argued that the test evolutions closely replicate the tasks firefighters actually perform on the job. Fire departments have found the public, unions, courts, and even candidates themselves less willing to dispute a test that, on its surface, resembles what people generally believe firefighters are required to do.

---

*The Americans with Disabilities Act provides for simulated or actual task demonstration in order to establish ability within the context of that Act.[2]

Job-simulation tests have numerous shortcomings. They often include tasks on which a person who has practiced or received training will perform better than an untrained one, which makes an objective assessment of the candidate's raw ability almost impossible. The tasks may also be unsafe for an untrained person to perform. In addition, job-simulation tests sometimes require the candidate to perform tasks alone that normally are performed by two or more firefighters.

Also, these tests often use scoring systems that require the candidate to perform the test at an "all-out" pace rather than a pace actually used on the fireground. Such scoring systems identify the fastest candidate, but this determination is meaningless in the absence of data to demonstrate that faster candidates actually make better firefighters. (Scoring systems are discussed later in this section.) Lastly, the events on a job-simulation test may be so unrelated to actual fireground tasks that they do not actually measure what they are intended to measure.

Fitness assessments. Some fire departments assess fitness rather than task performance when selecting firefighters. Their philosophy is that a physically fit candidate is most likely to succeed on the job. Fitness assessments may include such events as a single bench press with a designated minimum weight, a sit-and-reach flexibility event, push-ups, sit-ups, and dumbbell curls. In the past, scoring could be scaled to the candidate's age, body weight, and gender to correlate with general fitness standards for each group. (This would mean a 23-year-old man weighing 210 pounds might be required to bench press 140 pounds, while a 34-year-old woman weighing 150 pounds would be required to do only 80 pounds; in flexibility measures, women might be required to out-perform men in order to receive a passing score.) Such scaling may now be open to challenge under the Civil Rights Act of 1991. (*See sidebar.*)

The Civil Rights Act of 1991 and test scoring

The Civil Rights Act of 1991 states "It shall be an unlawful employment practice…to adjust the scores of, use different cut-off scores for, or otherwise alter the results of employment related tests on the basis of race, color, religion, sex, or national origin."*

The legislative history of this Act makes it clear the language was intended to prevent adjustment of written test scores based on the applicant's race. When subsequently the Act was broadened to include all employment tests and not just written ones, this language was not revisited. The legality of gender- and age-adjusted fitness test scores under this provision of the Act remains to be clarified.

*Title 42 U.S. Code. §2000-2 (L).

Fitness assessments are not widely used as entry-level tests; where they are, in most cases, elements from proxy or job-simulation tests also are included in the testing process. The primary advantages to this type of test are that it provides a comprehensive view of the candidate's overall fitness and does not rely on skill-dependent tasks for its measurements.

Criticisms of fitness assessments as entry-level tests usually focus on the scaling of scores to the candidate's age, gender, and weight. While this may be the accurate way to determine how a candidate's fitness level compares to that of others in his or her category, it runs counter to the performance approach that measures the candidate by the demands of the job. Demonstrating the criterion validity of such a test, should that become necessary, would appear to be difficult. Even if the scores are not scaled, it may be difficult to relate passing scores on the test to actual job demands. But where tests do not have a disparate impact, Title VII does not require the employer to demonstrate the test's validity or relevance to the job. (A test must be demonstrably job-related, however, if it tends to screen out individuals or groups protected under the Americans with Disabilities Act[3] or other civil rights statutes.)

No entry-level physical test. Some smaller fire departments have never used physical performance tests as part of their entry-level process, relying instead on training and the probationary period to weed out those who cannot perform the job. At least one large department in the early 1990's abandoned its entry-level physical test in favor of the screening provided by the training process.

Having no physical component to a firefighter hiring process other than a medical examination seems like a radical notion. The job does require strength and fitness, even if the exact levels of each are difficult to determine. The idea, however, has its advantages. It does not disqualify applicants on the basis of a test that may not be valid. It permits recruits to be trained before their abilities and performance are assessed. And it eliminates the expense involved in developing, validating, and defending a test.

It also has several disadvantages. It may put candidates into firefighter training who do not have the basic fitness necessary to complete the program. This places a burden on the training staff, and is unfair to the candidate who may have quit another job in the sincere belief that he or she had a good chance to become a firefighter. Not giving a physical test to firefighter candidates also may generate controversy and disharmony within the department.

A department that does not give an entry-level physical test must use other methods to sort qualified job applicants from unqualified ones. Typically, that process then happens during recruit training. The training must be of top quality, staffed by skilled and knowledgeable instructors and evaluators. If it is not, it cannot function well as a selection process. The need for a stringent medical physical (within the limits of the Americans with Disabilities Act) becomes more critical, particularly one that assesses the cardiovascular system and the potential for back injury.

Modified approaches. An alternative to having no physical performance test is to offer the fitness assessment or physical performance test to applicants on a voluntary basis. Candidates take the test as a self-assessment, to measure their level of performance against what will be required of them on the job. This gives them the opportunity to improve on their fitness levels or withdraw from the hiring process, if necessary.

At least one major fire department uses a two-tiered approach. It gives applicants a physical performance test--the same test all fire department personnel take every year--but allows them several minutes longer to perform it. By the end of their recruit training, firefighters must perform up to the incumbent standard.

Designing physical tests

A valid physical performance test gives the fire department a way to predict the posttraining level of job performance of a group of currently untrained people. In other words, it accurately identifies those job applicants who, with training, will be able to become successful firefighters. It does so by assessing only factors related to the job, and is not affected by other factors such as a candidate's race, gender, age, or size. An entry-level physical test must be based on a job task analysis done in the department using the test. It also should be economical and relatively easy to administer.

Fire department managers should look critically at their department's physical test to see if it truly measures what it is supposed to measure, in the fairest way possible. In particular, the test should be comprehensive, should contain few if any elements at which one improves with training, and should reflect actual fireground practice.

Comprehensiveness. Many entry-level firefighter tests primarily assess speed (anaerobic capacity) and upper-body strength. But a firefighter's job takes more than speed and strong arms. A comprehensive test will assess the entire range of physical abilities needed on the job.

The *Uniform Guidelines** require "the behavior(s) demonstrated in the selection procedure (to be) a representative sample of the behavior(s) of the job in question…"[5] If, as often happens, women compare favorably with men on the criteria that are not tested, and candidates are tested only on those areas where men tend to outperform women, the results will inaccurately show women to be less qualified for the job than a more comprehensive test would indicate. A test that overemphasizes certain abilities and neglects others required by the job is unlikely to be found valid.

Test elements on which training improves performance. The Equal Employment Opportunity Commission (EEOC) *Guidelines* specifically discourage skill-dependent tasks in entry-level tests, warning against testing on the basis of "skills that can be learned in a brief orientation period." Such elements should be minimized on an entry-level test; the tasks selected should be those that require the least training and the least knowledge of test equipment to perform.

An entry-level test should not attempt to measure an untrained candidate's ability to perform firefighting skills. Entry level is, after all, the bottom of the ladder. It's where a fire department brings raw recruits into the employee-development process. If you test people on trainable skills, you will end up selecting people who have had training, not those who necessarily have the basic abilities you want. If entry-level firefighters will all receive firefighter training before going on the line, the test should not inadvertently give preference to candidates who have prior experience. (You may choose to favor those with experience by giving extra points through the interview process or elsewhere, but that is not the function of the physical performance test.) Holding test practice sessions (*see "Recruitment" section, page 2*) can reduce the impact of skill-dependent elements on a test, but it is even more effective to eliminate these tasks wherever possible.

Reflection of actual fireground practice. In its pace, time limits, technique, and the selection and sequence of tasks, an entry-level test should reflect how tasks are done on the fireground.

Time limits. A test that requires successful candidates to perform at an all-out pace does not represent the way the job actually is done. While many fire departments set realistic time limits for their tests and specify that a walking pace be maintained throughout, others use tests that, by design or unintentionally, require an all-out, anaerobic pace. Although actual fireground performance calls on a firefighter's aerobic capacity, these tests require speed-to-completion performance and therefore, in the way they are performed by successful candidates, measure only anaerobic capacity. Their pace and intensity do not match the way firefighter tasks actually are done.**   (*See further discussion under "Scoring systems," below.*) If even a brief rest period breaks up a test designed to measure aerobic capacity, stamina, and endurance, it may instead become two or more anaerobic or speed tests.

Some sequential tests lump all tasks together under a single overall time limit instead of timing each task separately. This does not measure minimal ability in any single task. A candidate could perform one task at a speed that could be unacceptably slow on the job, but still pass the test by completing other tasks at speeds that might be unnecessary, irrelevant, or even dangerous on the job.

---

*The *Uniform Guidelines* on *Employee Selection Procedures* issued by the Equal Employment Opportunity Commission (EEOC) define the law on employment testing, under Title VII of the Civil Rights Act of 1964.

**In 1989, New York City administered an intense, all-out test scored on speed-to-completion. One male candidate died during the test, and more than 30 other men were hospitalized or treated for renal failure. Other cities that have administered comparable tests have seen similar medical problems result.

Technique. Different firefighters perform fireground tasks in different ways. This means the techniques used to accomplish a task on the job will vary depending on who is in the workforce. Firefighters who have a lower center of gravity or greater strength in the lower part of their bodies may find it more effective to use a different technique for pulling a ceiling or cutting a roof than those who have greater strength in the upper part of their bodies.

Therefore, events on a job-simulation test should not be based on a single technique for a fireground task. For example, a test might require candidates to use the same technique to lift a dumbbell that larger men use to lift ladders on the job. Since other equally good ways to lift ladders exist that women may use more effectively, the dumbbell lift could eliminate women candidates who could actually perform well on the fireground. If a more diverse workforce alters job skill performance by using a variety of techniques for lifting ladders, the correlation between the exam and fireground performance can decrease significantly.

Test events that are simulated in order to reduce skill dependence can still reflect the variables of fireground practice and individual technique. One example is a debris-removal evolution that uses two buckets and a shovel. Candidates are required to simulate the removal of debris from a building by moving a certain weight of gravel from Point A to Point B. They may use one or both buckets; they may choose to put only a moderate amount of gravel in the buckets each time and make more trips, or make the buckets heavier and carry them fewer times. Each candidate is free to modify the task to his or her own abilities, while still achieving the overall goal of moving the gravel within a designated time.

Sequence of events. For job-simulation tests, the sequence of events should reflect the order in which the tasks are done on the fireground. On the job, women firefighters often use their lower bodies as well as their upper-body strength in forcing entry. An arbitrary sequence of tasks on the test, however, might exhaust the lower-body muscles by the time the candidate reaches the forcible-entry simulation (which normally would occur early at a fire scene). Candidates then will have to perform the task in a way they might not actually perform it on the job: by using their arms only. The test then will measure the candidate's ability to perform forcible entry with his or her arms, and not the ability to accomplish the task in the way the person would normally do it.

Scoring systems

The scoring system used to rank candidates is a crucial factor in determining the fairness of a test.[6] A bad scoring system can negate the validity of an otherwise valid test, while a carefully chosen scoring system may help counteract the negative impact of an imprecisely designed test. Entry-level physical tests for firefighter candidates may be scored in any one of a number of ways; following are some of the options and their pros and cons.

Rank-ordered, speed to completion. Under this type of scoring, candidates are encouraged to go through the test at the maximum possible speed. This encouragement is either overt (when candidates are instructed or cheered on to perform as quickly as possible) or implicit (when candidates know those with the fastest times will receive the highest scores and have a better chance of being hired).

It is important to address the underlying assumption of such scoring systems: that the person who gets through the test fastest will be the best firefighter. Most people would agree that generalized attributes such as strength, fitness, flexibility, and endurance are good things for firefighters to have. It is less obvious that extreme levels of these attributes are not necessarily relevant to the job.

To draw a parallel, if you needed a car that would perform efficiently at 65 mph, and a Ford Escort or Honda Prelude suited your needs, a Ferrari wouldn't necessarily be even better. If you needed a family car to haul kids, pets, and groceries, a moving van wouldn't necessarily be better than a station wagon just because it would haul even more. Similarly, someone who can bench-press 500 pounds or run a 4-minute mile will not necessarily be twice as good a firefighter as someone who can only bench-press 250 pounds or run a mile in 8 minutes. Many factors come into play in the makeup of a good firefighter. Overemphasizing any aspects, such as strength and speed, will result in other important characteristics being overlooked.

Rank-ordered, speed-to-completion scoring is most commonly used on job-simulation tests. Because these tests **look** like firefighting, it is easy to think the fastest candidate will be the best firefighter. But such tests are not really **done** like firefighting. They offer only an approximate representation of some firefighting tasks without the overlays of the real fireground, where obstacles such as darkness, ice and snow, debris, constricted spaces, noise, smoke, poor communications, and general confusion all constrain performance. One exercise physiologist has commented

> There may be a distinct difference between the levels of physiologic fitness and performance required to effectively perform a job such as firefighter compared to the requirements to excel on a timed, all-out physical test comprised of specific tasks…The testing conditions under which the (test) is performed are so different from the dangerous, dark and smoky conditions of actual firefighting that extreme speed on the (test) may indicate speed beyond the amount which can be used during actual firefighting.[7]

In addition, most fireground tasks actually are carried out by two or more people, while testing can normally be done only on individuals.

There is another crucial difference between job-simulation tests and real firefighting. Test-takers know in advance exactly what tasks they must do and when they will be able to quit. Most tests take only a few minutes to complete, and the candidate's condition at the end is not taken into account. If he or she collapses and cannot function further without a rest, or even needs medical attention, this does not affect his or her score. Candidates can thus push themselves to an all-out pace they could never achieve or sustain on the fireground. Firefighters at a fire are typically called on to work for 20 minutes to an hour or even longer, at a varying and unpredictable series of tasks, without the opportunity to rest.

Faster-is-better scoring invites higher scrutiny of a test's validity. Any test must select accurately, from among untrained candidates, those who will be able to complete firefighter training successfully. A rank-ordered, speed-to-completion test also must correlate specifically the candidate's test score with his or her level post-training job performance. The Federal law on employment testing states:

> Evidence which may be sufficient to support the use of a (test) on a pass/fail basis may be insufficient to support the use of the same procedure on a ranking basis.[8]

Few if any tests are so accurate that the required correlations can be demonstrated reliably. If they cannot, the fire department or other testing agency cannot legally justify scoring its test in this way.

Speed-to-completion scoring encourages unsafe behavior on the test. This can result in injury or even, as in one case, death for the candidate. As this unsafe behavior may be rewarded by a job offer, the fire department thus may select candidates who exhibit behavior it does not, in fact, condone. Many other factors affect candidates' job performance besides how fast they can get through one set of tasks on a given day. If this were not true, the fast new recruit always would be more valuable to the fire department than the somewhat slower veteran.

One fire service manager has openly questioned whether there might in fact be a direct correlation between high scores on the entry-level test and later incidence of job-related injuries.

Whether this is due to exertion beyond reasonable limits, ego-driven recklessness or a combination of…factors, extraordinary performance on agility tests does not appear to be a valid means by which to predict career performance or longevity.[9]

Banding. This modification of rank-ordering groups candidates according to their scores: for example, everyone scoring from 90 percent to 100 percent might be in the top group or "band." All candidates in each band are considered to be equally qualified. This option can pose the same problems as rank ordering if the bands are very narrow, since using bands that are only two or three percentage points apart is little different from rank-ordering. Using wide bands can alleviate some of the above-mentioned problems, but the system still can be manipulated to exclude minority candidates (women and people of color). When the bands are wide, it is unlikely anyone outside the top band will be hired. If that top band happens by accident or design to include only white men, only white men will be hired. In addition, wide bands dilute the pool of minority candidates who score well on the test. If hiring from the top band is done randomly, such as by lottery, a high-scoring minority candidate's chances of being hired are reduced.

Pass-fail. Many fire departments have found a "qualifying," or pass-fail, scoring system best suits their needs. In this approach, the physical performance test is viewed as a screening device to separate qualified from unqualified candidates, and all who pass it are viewed as capable of becoming firefighters. The cut-off score is set in advance, based on the test times of a representative sample of men and women incumbents who are satisfactory employees of the department.

Pass-fail scoring recognizes that physical performance tests are inherently imprecise and imperfect. It allows candidates who have a reasonable expectation of success to continue through the hiring process, without trying to draw fine lines based on a few seconds of speed on one set of tasks on a particular day.

Continuous-performance. One fire department has implemented a new departure in physical performance test administration. Tests that require the candidate to complete a specified set of tasks in a designated time have two weaknesses in particular: often they do not assess cardiovascular fitness, and they do not consider the candidate's physical condition at the end of the test. Because both of these are of significant concern to fire departments, the continuous-performance option was developed. Candidates are required to perform the tasks in a continuous circuit for a specified period of time (such as 20 minutes) linked to what is expected of a firefighter at an actual fire. A minimum number of circuit or task completions is prescribed, based on incumbent firefighter performance. This option screens out those who are able to rush through a set of tasks quickly but do not have the endurance to continue performing the way they would need to on the fireground.

Fire departments and other agencies with testing responsibilities must choose carefully when establishing a scoring system for their entry-level physical performance test. The "more is better" approach of rank-ordering or narrow banding has a strong superficial appeal, and changing from rank-ordered scoring to a pass-fail system can be controversial. Nonetheless, fire service managers' responsibility for test validation includes not only the selection of tasks and how they are to be performed, but also the scoring system used. A flawed scoring system can result in the hiring of less qualified candidates or the arbitrary rejection of more qualified ones: exactly the opposite of what a test is supposed to do.

Test administration

A physical performance test can accomplish its intended purpose only if it is properly administered. The guidelines below will help administrators avoid some common pitfalls in giving entry-level tests to firefighter candidates.

- Design the test, and set the pass-fail point, before the application period is announced. Firefighter recruitment literature should offer a complete description of the test, how it will be scored, and how the final hiring list will be established.

  Some fire departments have been reluctant to give out information about scoring systems, cut-off times, or even about the composition of the test itself. This affects the test performance of those who require more preparation time. A candidate's willingness to spend time and effort preparing for a test is a strong indicator of his or her commitment to the job. This should be viewed positively, not as a strike against him or her.

- If candidates are to wear protective gear during the test, make sure it is available in sizes that will fit everyone. Qualified applicants may be eliminated if they are forced to wear gloves or boots that are too large, or a poorly fitting self-contained breathing apparatus (SCBA) that shifts and throws them off balance. These factors primarily affect women. Many departments substitute an appropriately weighted vest for the turnout coat and SCBA. These vests are adjustable to a wide range of sizes and do not give an advantage to one group over another, as the use of firefighting gear may.

- Carefully check the equipment to be used on the test to be sure it functions as well as equipment used in the field. For example, one fire department's test included a ladder extension. The rope on the test ladder was of a smaller diameter than that on ladders used on the fireground, making it much more difficult to grip effectively. Making a task less difficult on the test than it is in the field is an option, to compensate for the fact that recruit firefighters will receive training and practice that will improve their skills. It is difficult, however, to justify a task on the test being **more** difficult than it would actually be on the job.

- All performance techniques that will be permitted on the test should be demonstrated to the candidates. These should be the same techniques that have been taught at test practice sessions. Personnel who assist at test practice sessions should be knowledgeable about the test events and helpful in suggesting safe, effective techniques to all candidates. An overlap of personnel between the practice sessions and the test itself will provide consistency in techniques and other information.

The limits of validity

A good physical performance test will do more than just meet the standards of the law. A test may comply with Federal law and still not necessarily select the best candidates. This is due partly to limitations of the *Uniform Guidelines'* approach to testing, and partly to the impact of past and current discriminatory practices. Specific problems include flawed job task analyses, inaccurate determinations of a test's predictive value, confusing common abilities with needed ones, and errors in setting cutoff scores.

Failures of the survey process. In order to establish a fire department's entry-level physical performance standard for firefighters, you must assess what it is firefighters actually do and how they do it. This is called a job task analysis. It should provide a picture of the job of firefighter, breaking duties down into their simplest components.[10] The analysis has an objective part--a list of tasks and how often they are performed--and a subjective view of how critical each task is. The test developer gathers information about the job from people on the job. The more accurate this information is, the better the job analysis will be.

The surveys used to determine critical job skills do not always work as intended. People already on the job and the evaluators themselves bring biases to the evaluation process. It is natural for a dominant, incumbent group to rate the traits they themselves hold in relative abundance as critical to the job, and to underrate the importance of traits they do not possess. Similarly, surveys written by evaluators who have preconceptions about the job often will reflect those preconceptions in subtle or obvious ways and affect the resulting data.

These biases produce inaccuracies that the surveys generally are not designed to eliminate, which allows values to be established by the dominant group instead of being created impartially.[11]

Inaccurate determination of a test's value in predicting job performance. Sex discrimination on the job can make an entry-level test that discriminates against women appear to have criterion validity. If women score lower than men on an entry-level test (but still high enough to be hired) and then suffer on the job from sexual harassment, poor training, or other behavior that diminishes their productivity or has a negative effect impact on their performance evaluations, a validation study will seem to show that poor performance by women on the entry-level test predicted their poor performance on the job.*

Confusing common abilities with needed ones. Incumbents may do well on tests that do not measure abilities required on the job. A test that only measures common traits (ones many people on the job hold) instead of needed skills (ones the job demands) may appear on its surface to be valid. As an example, imagine that an old, flawed test had given an advantage to right-handed candidates, so that no left-handed firefighters were on the job. A profile of the department then could be used to validate a new test that would result in the selection of future generations of only right-handed firefighters. Make sure your test does not select for irrelevant traits: those possessed by the majority of firefighters because the firefighters happen to be male, as opposed to those that are important to task performance.

Errors in determining test cut-off scores. The process of determining the cutoff point for a passing score of work-performance tests is also subject to error. Some fire departments norm their test under race-like conditions that artificially enhance the value of speed, instead of mandating a normal fireground pace. Even where speed is not encouraged specifically, incumbents tend to speed up their performance for several reasons. Out of interpersonal or intercompany rivalry, firefighters may perform quickly in order to try to beat each other's times. In other cases, firefighters may fear that those who perform more slowly will suffer repercussions from management, and they thus push themselves to go through the test quickly. Finally, if the test is normed in an atmosphere that has been heavily polarized, such as by a lawsuit and the reaction to it, incumbents may deliberately perform the test as quickly as possible in order to distort the passing time and make the test more difficult for candidates.

Physical performance testing of incumbent firefighters

Like entry-level testing, the widespread use of periodic physical performance tests to determine the "fitness for duty" of incumbent personnel is a relatively recent phenomenon. A number of national and local standards, including NFPA 1500, require emergency responders to be evaluated and certified as meeting their departments' physical performance requirements on an annual basis. The NFPA standard does not specify the level of performance required. This responsibility, along with the challenge of designing a relevant, reliable, and valid test, is left to the local department.

Physical performance testing of incumbent firefighters often is controversial, especially when performance below the level set by the department is met with punitive measures. Mandatory testing of this type has resulted in numerous firefighter injuries and even deaths,[13] and may create a stressful environment detrimental to the morale of the whole organization.

---

*In 1988, Columbia University surveyed attitudes of male firefighters and officers towards women firefighters on the New York City Fire Department. Men of all levels and experience rated FDNY women firefighters (all of whom had been on the department for 5 years) as less competent than brand-new male recruits on every firefighting task except skill in community relations. For tasks requiring strength--carrying hose to the fifth floor, making a rescue--men saw the gap in "ability" between new recruits and women as being even wider in favor of the male recruits. Women's interpersonal relations skills were rated significantly lower than new recruits if the ability to take hazing and pranks was the criterion. Men who actually had worked with women firefighters rated them slightly higher across all criteria than did those who had not.[12]

Physical performance tests are not a sufficient guarantee that responders are fit or capable of doing their jobs. If employee health, safety, and fitness, and a reduction of occupational injuries are the goals, a comprehensive wellness program is the appropriate choice. The wellness approach emphasizes prevention, education, and a balanced health and fitness program. Wellness programs use medical exams and fitness testing to identify at-risk firefighters. Physical performance testing may be used to diagnose deficiencies or problem areas and help the firefighter achieve the desired level of performance.

Fire department managers should exercise caution when adopting a physical performance testing policy for incumbent emergency response personnel. Periodic testing should not replace regular evaluations of the onscene performance of all firefighters. These evaluations help determine whether substandard job performance is related to shortcomings such as lack of training and practice or poor communications. Once the problem is diagnosed, the department can intervene appropriately.

Whatever policy is adopted, it should be recognized that physical performance testing alone cannot ensure that firefighters are healthy, fit, and prepared to respond to emergencies. Measuring performance once a year may provide useful information, but it is not a substitute for training, education, and regular exercise.

Conclusion

Fire department managers should examine entry-level testing processes carefully for elements that give any group an unwarranted advantage. Bias is not always obvious. It may be concealed in "substitute indicators" such as using speed as the primary scoring criterion. It may be hidden in the job task analysis, or it may be caused by test events that do not measure what they are designed to measure because they allow only one technique in the performance of a task, or incorporate elements on which candidates can improve with training. The cause also can be as simple as candidates being required to wear firefighting gloves on the test, when the gloves do not fit all of the candidates.

Fire departments that use job-simulation tests to assess candidates' performance have found these "look-alike" tests to be widely accepted. Some exercise physiologists have developed job standards based on complex measures of the energy costs associated with fire suppression tasks, and claim their criterion-linked standards are more defensible in court. Whatever the type of test used, when it has disparate impact on a protected group, the test may be challenged in court, and the fire department using the test bears the ultimate responsibility of defending it.

Many fire chiefs have found establishing the validity of a physical test to be lengthy, difficult, and expensive. The employer must justify all aspects of the test, from the selection of tasks to be performed to the time allowed for completing the tasks. Regardless of who wins in court, the enormous expenses of time, money, and morale can produce a no-win situation for both sides in a legal challenge.[14]

Notes:
[1] *Berkman v. Koch*, 536 F. Supp. 177 (E.D.N.Y. 1982), aff'd 705 F.2d 584 (2d Cir. 1983).
[2] 29 C.F.R. §1630.
[3] 29 C.F.R. §1630.14(a).
[4] 42 U.S.C §2000e *et seq*, as amended; 29 C.F.R. §1607: *Uniform Guidelines on Employee Selection Procedures.*
[5] 29 CFR. §1607.14 (C)(4).
[6] The *Uniform Guidelines* address cut-off scores at 29 C.F.R. §1607.5(H).
[7] McArdle, William; St. Paul deposition, 1990.
[8] 29 C.F.R. §1607.5(H).
[9] Osby, Robert. "Guidelines for Effective Fire Service Affirmative Action," *Fire Chief*, September 1991, p. 53.
[10] 29 C.F.R. §1630.14(a).
[11] See *Kilgo v. Bowman Transp.*, 570 F. Supp. 1509 (N.D. Ga. 1983), aff'd 789 F. 2d 859 (11th Cir. 1986). Job analysis found wanting for several reasons, including the fact that it profiled the trucking industry according to its present over-representation of male employees. See also *Vulcan Pioneers v. N.J. Dept. of Civil Service*, 832 F. 2d 811 (3d Cir. 1987).

[12] Center for Social Policy and Practice in the Workplace, "Gender Integration in the New York Fire Department: A Review and Recommendations," Columbia University School of Social Work, 1988.

[13] *International Fire Fighter*, Vol. 79, no. 1, January-February 1995, p. 1.

[14] The philosophical basis for part of this discussion was based in part on an excellent analysis of testing and EEO by Kellman, M., "Concepts of Discrimination in 'General Ability' Job Testing,' 104 *Harvard Law Review* 1158 (April 1991).

Legal aspects of physical performance testing

Validating entry-level tests

Test validation is a confusing issue for many people. The word "valid" itself is highly misunderstood and misused. The misunderstanding arises in part because "validity" has a specific legal definition in this context and does not simply refer to a general sense that the test is job-related or contains elements that look like things firefighters do. In addition, fire service administrators and the developers and marketers of tests sometimes use the terms loosely to imply that a particular test is guaranteed to withstand any legal challenge.

Why is the legal definition of test validity a fire service concern? Because it is the primary way an employer can defend itself against claims of discrimination in hiring. If someone files a complaint that a fire department's entry-level test is unfairly discriminatory against women, African-Americans, or people of a particular religion, the fire department must be able to show its test is valid.[1] If the test is not valid, the complaint is likely to be upheld.

The legal definition of "validity" is found in the Equal Employment Opportunity Commission (EEOC)'s *Uniform Guidelines on Employee Selection Procedures*[2], which in turn refer to the 1985 publication, *Standards for Educational and Psychological Testing*, of the American Psychological Association (APA). This document defines validity as "the degree to which a certain inference from a test is appropriate or meaningful."[3] When used in hiring, a test is valid if decisions made on the basis of test scores correspond to eventual job performance. In other words, a person's test performance should be a reliable forecaster of her or his job performance.

How does one determine if a test is a reliable predictor of performance and will therefore stand up to such challenges? The EEOC has adopted three methodologies of the industrial psychologist to assess a test's job-relatedness. These are referred to as "criterion," "content," and "construct" validity. No one validation method is recognized as superior; the choice of method for demonstrating the job-relatedness of a test depends on the feasibility of conducting the research to document it.

A test with **criterion validity** accurately predicts, or is significantly correlated with, employees' actual work proficiency. Criterion validity compares test results with actual job performance. It is generally used to validate tests that claim to measure aptitudes that will be turned into specific skills in the work situation. To establish criterion validity in the fire service, a test usually is administered to a group of employees on the job, and their test scores are compared with their current work performance. (This is referred to as "concurrent" criterion validity.) A weakness of such studies is that different levels of experience may muddy the correlation if the study doesn't take into account variables among the incumbents such as time on the job and the range of job assignments.

A test with **content validity** duplicates or represents actual job duties: it compares an applicant's current skills and abilities with specific job functions. The *Uniform Guidelines* specify that the skills and abilities tested must be "representative" of those required on the job. The knowledge, skill, or ability must be "used in and necessarily prerequisite to performance of critical or important work behaviors."[4] Content validity studies rely above all on a thorough job analysis. Often this type of validity is viewed as inappropriate for inherent aptitudes and for special skills learned on the job.

A test with **construct validity** identifies the traits required for successful job performance. Construct validity compares more abstract and indirect qualities (for example, problem solving ability) with particular job duties to which the qualities are relevant. Arguably, construct validity is the more comprehensive form of validity, because it incorporates methods of both criterion and content validity. In the past, construct validity has been assumed to be useful primarily for testing for mental or psychological traits.

31

In order to establish a test's validity in any of these three ways, the employer must conduct a thorough job task analysis. This analysis will identify what kinds of abilities are required by the job and decide what must be tested for. In addition, each validation method has specific requirements:

- **Criterion validation** requires an existing measure of job performance so the correlation between test results and job performance can be examined.

- **Content-validity** studies must do more than prove mere "facial validity," i.e., that the tasks the test-taker is being required to perform **look** the same as tasks done on the job. A content-validity study also must investigate what skills will be developed readily on the job, since it would be inappropriate to test for those skills at the entry level or before sufficient training had been given.

- **Construct-validity** studies do not directly measure actual job tasks; they use proxy measurements to determine if the applicant has the necessary job abilities, such as creativity, which cannot be measured directly.

The *Uniform Guidelines* define the job task analysis as a detailed statement of work behaviors and other information relevant to the job. It also has been defined as a procedure undertaken to understand job duties and behaviors and performance standards.[5] Job analysis procedures also gather information about how the work is accomplished, the setting in which it is performed, and the resources and tools used. Test results can not be correlated with actual job performance unless it has been established what "good" job performance is. Therefore, the analysis should be a thorough survey of the relative importance of various skills involved in the job and the degree of competency required. The sources providing the job analysis must be credible and unbiased.[6]

When someone claims a particular fire department's test is "valid," they may, in fact, only mean that a job task analysis has been done. They also may mean that the department hired a statistician (or an industrial psychologist, or some combination of testing experts) to prove that the test has one or more of the three kinds of validity. But a test truly cannot be said to be valid until it has been challenged and, after a fact-finding process, been found to meet the requirements of the EEOC *Guidelines* and applicable State and local law.

Even after a test has been found to be valid for a particular employer, it may not be valid for another employer. The second employer must still conduct its own job task analysis and conduct a transportability study showing that the second employer's job is essentially the same as that of the first employer. Local conditions prevail in the fire service workplace; thus, no test can be validated on a national basis.

Challenges to employment tests often become an expensive battle of experts. Job task analysis has become a highly technical art. For these reasons alone, fire departments are advised to spend time and money up front, before incorporating any new test into the hiring or incumbent-testing processes.

Legal considerations affecting scoring systems

Employers scoring an entry-level test on a pass/fail basis must fulfill two requirements in justifying the cutoff score. First, the test must be "reliable": that is, if it were graded, or the candidate were evaluated, by two different people, the results would be fairly similar. Second, the employer must have a "justifiable reason" for adopting the score.[8] If a test is used to rank applicants, the employer must show that higher scores correlate with better job performance. The Supreme Court has held that where the disparity of impact is greater at high passing scores than low, any alleged correlation between higher scores and better job performance must be "closely scrutinized."[9]

Can a fire department justifiably require job applicants to perform tasks (or perform at speeds) that firefighters on the job cannot themselves perform? The rationale for rank-ordering and for high cutoff scores for pass-fail tests traditionally has been based, in part, on common assumptions about the effects of age. Fire departments have argued that because workers' physical abilities "naturally" deteriorate with age, employers are justified in requiring greater physical abilities from new hires than from incumbent firefighters.

This assumption is becoming increasingly questionable. The EEOC in 1992 issued findings arguing that Congress should revoke the fire service's exemption from the Age Discrimination in Employment Act. The study that was done to research the issue concluded that age was not a reasonable predictor of employee performance in the fire service: that "accumulated deficits" in firefighter physical ability were only marginally associated with age.[9] As attitudes change on the issues of diet, exercise, and smoking, it will be increasingly difficult to argue that the fire service can maintain separate standards of physical fitness for new hires and incumbents. Rather, the trend will undoubtedly be to have one fitness standard for all firefighters.

Notes:

[1] In defining whether a test has "adverse" or "disparate" impact--thereby shifting the burden of proof to the employer to produce a business justification for the test--statistical evidence has come to dominate discrimination litigation. The EEOC's *Uniform Guidelines on Employee Selection Procedures* at 29 C.F.R. §1607.4(D) introduces some quantitative criteria by which to judge the significance of differential impact: the "four-fifths rule." This rule compares the success rate for a protected group with the rate achieved by the nonprotected group (e.g., white men) by way of a ratio. If the plaintiff shows that the success rate of the protected group is less than four-fifths (80 percent) that of the unprotected group, then the burden shifts to the employer to show that the disparity is produced by some factor other than discrimination. One obvious problem with the use of this statistical approach occurs when groups are being compared are small in numbers. Some courts have tended to disregard the statistical evidence of success rates as having no predictive value when the absolute numbers involved are low (i.e., two women and 47 men took the test, and neither woman passed it).

[2] 29 C.F.R. §1607.

[3] At p. 94.

[4] 29 C.F.R. §1607.14(C)(4).

[5] The American Psychological Association's Society for Industrial and Organizational Psychology's publication *Principles for the Validation and Use of Personnel Selection Procedures* (3rd ed. 1987).

[6] An early case defined the components of an "adequate" job analysis as establishing:

1) the competency of the person/entity preparing the job analysis;

2) the proficiency of the information-gathering phase of the study: sample size, data compilation (were data ignored?), the "casualness" of the approach, and whether confusing survey questions were used; and

3) the competency of the person/entity preparing the actual examination.

[7] In *Gillespie v. Wisconsin*, the court said the employer must use a professional estimate of the requisite abilities or at least a logical "breakpoint." 771 F.2d 1035 (7th Cir. 1985).

[8] Landy, Frank J. "Research on the use of fitness tests for police and firefighting jobs," The Pennsylvania State University Center for Applied Behavioral Sciences, 1992.

[9] *Guardians Assn. v. Civil Service Commission*, 630 F.2d 79 (2d Cir. 1980). Also see *Pina v. City of East Providence*, 492 F. Supp. 1240 (D.R.I. 1980); *Brunet v. City of Columbus*, 642 F. Supp. 1214 (S.D. Ohio 1986), *appeal dismissed*, 826 F. 2d 1062 (6th Cir. 1987), *cert denied*, 485 U.S. 1034 (1988); *Zamlen v. City of Cleveland*, 686 F. Supp. 631 (N.D. Ohio 1988), *aff'd*, 906 F.2d 209 (6th Cir. 1990), *cert denied*, 111 S. Ct. 1388 (1991).

Fire chiefs and physical performance testing

At a 1995 workshop on physical performance testing, Chief John Rukavina of the Asheville, North Carolina, Fire Department offered his perspective on why many fire chiefs prefer "package" physical performance tests (pre-existing tests that can be bought or copied). He also provided some questions that should be asked when a fire department or other testing entity is considering a new physical test. That information is summarized here.

Why are "package" physical performance tests so popular with fire chiefs across the U.S.?

- Fire chiefs generally don't have the resources or time to conduct a proper job analysis. (While the Americans with Disabilities Act requires a general physical ability analysis, this analysis is done in terms of what physical abilities are required--not how much or how fast.

- Fire department budgets may not have prioritized funds to spend on the expertise required to conduct a suitable job analysis.

- Fire chiefs sometimes look for a test that is credible to other firefighters and to the fire chief. Tests that "look like firefighting" may seem more credible than a more valid test that does not.

- Fire chiefs often prefer to use someone else's test. This is a variation of the "white-out" approach to management, in which a fire department adopts another department's regulations or SOP's by whiting out the original department's name and inserting the borrowing department's name. It's based on a lack of resources, time, or expertise available to create and validate a local physical ability test.

- Fire chiefs may not want to invest a lot of money in the physical performance test itself.

- Some fire chiefs like physical performance test scores that remove any responsibility for judgment on the chief's part. Chiefs often prefer tests to be scored on a rank-ordered basis, leaving them no alternative but to hire the person at the top of the list, rather than being able to use their judgment and then have to defend their decisions. Chiefs may prefer not to have any discretion where using it will leave them open to criticism.

Some "consumer questions" to ask when considering a new physical abilities test:

- On what is the physical abilities test based? Is it based on what the department has identified as "successful performance criteria?" What are the department's criteria for performance success on the fire scene? Which of these criteria is this test a proxy for--that is, which ones is it intended to evaluate? Are those criteria available for review?

- What weight does the proposed test give to effectiveness? Safety? Speed? How would your department rank the importance of these three qualities to actual fireground performance? Is the relationship between the weights given to these factors on the test consistent with the weight given these factors on the fireground?

- If you are using this test as a performance standard, how do you take into account the fact that firefighters perform in teams on the job?

- Will incumbent firefighters have a preparation period in order to condition to meet test standards? What will happen to an incumbent firefighter who doesn't pass the test? What will you do if 30 percent of your incumbent firefighters fail the test?

- Does the test have a disparate impact on firefighters over 40?

- Will all personnel with fireground responsibilities, including the fire chief, be required to take the test?

Firefighter training

Firefighter training in most departments in the United States is desperately in need of an overhaul. Training should be the cutting edge of the fire service, keeping firefighters and officers skilled and current as it leads fire departments into the future. Instead, it is often underprioritized, underfunded, disrespected, or ignored entirely. The fire department and the fire service as a whole suffer as a result.

Training and women firefighters

Poor training affects women firefighters in particular, in several ways. Female recruits and firefighters are often subject to higher scrutiny or stricter standards than their male counterparts. Many feel they are "under the microscope": that their mistakes are noticed more readily or taken more seriously, and that they must perform much better than their male counterparts in order to be thought nearly as competent. Such a high level of performance is difficult to achieve when training is inadequate.

Firefighter training often fails to address individual differences in firefighters' backgrounds and learning styles. Typically, firefighter training is modeled on methods used in team sports and the military. These can be effective with recruits who have sports or military backgrounds, or who learn well from these methods. Often it is ineffective with those who don't, which includes many women. As one woman firefighter said, "I don't think women respond as men do to 'scream in your face' mentality of teaching." Another said:

> I am not a hands-on person 'who just has to see it once.' I need to read and think about things and talk with people to gain understanding. In the fire service, this is often considered to be questioning authority.[1]

Training often fails to address other differences among firefighters, particularly new recruits. Women who enter firefighter training often have strong academic backgrounds and excellent skills in many areas that do not include mechanical knowledge or familiarity with tools. Firefighter training, traditionally aimed at male students who have learned basic mechanical skills elsewhere and are familiar with many tools, typically does not address the learning needs of others. One firefighter described her experiences:

> When I was first being taught how to open a hydrant, they handed me the hydrant wrench and told me, "You use this just like you would any other wrench." I had never, to my absolutely certain knowledge, ever held a wrench of any kind before. They tell me to cut along the floor joists; I have never wielded an axe; never heard the term "joist."[2]

On the other hand, strong academic backgrounds and other nontraditional skills of new recruits can enhance and expedite the training process in those areas. Teaching hydraulics, for example, is often much easier with college-educated recruits than it is with those who barely passed high-school math. As the fire service workforce continues to diversify, instructors are training more recruits, of both sexes, whose backgrounds differ from that of the students for whom the training was originally designed. Such instructors find it appropriate and effective to adjust their training programs accordingly.

Inadequate training also affects women more than men because of cultural differences between the sexes. Typically, men are more willing than women to function confidently in a situation with a minimal amount of knowledge; women require a more secure learning base to provide the same level of confidence. Poor training will thus leave women feeling less competent than their male peers. Women are also more willing than men, in general, to admit they don't know something and to ask questions in order to learn. Instructors used to training only men and unaware of these differences often view the questions or lack of confidence as incompetence and evaluate the women negatively as a result.

Changing the way we train

The purposes of fire service training should be

- to teach new fire recruits what they need to know in order to function well on the job;
- to keep skills current for all fire personnel;
- to introduce new ideas, techniques, skills, and equipment to all personnel; and
- to maintain and improve the function of individual crews, of groups of crews working together, and of department personnel working with personnel from other agencies.

In practice, fire departments' training programs rarely meet all these goals. To complicate matters, other training agendas and goals are sometimes included or substituted, such as:

- to intimidate recruits, "see who can cut it," and fire some of them;
- to punish incumbent firefighters; or
- to go through the motions so the department looks good on paper.

For the good of the entire department, as well as to help ensure that women firefighters operate on a level playing field with men, fire service leaders should reevaluate their department's training. Some questions to consider are

- Are the people who coordinate, design, and deliver training chosen for their abilities in these areas and are they good at their jobs?

- Does the training curriculum address all of the department's current training needs, with realistic priorities?

- Are training and teaching methods flexible, to reach as wide a range of students as possible?

- Is the training environment positive and supportive, rather than punitive or intimidating?

- Are training facilities and equipment adequate and current?

These questions will be discussed below, in the context of both recruit and incumbent firefighter training.

Recruit training

The firefighter academy, or basic recruit training, is the first extended point of contact between the fire department and its new personnel. Recruit training is where the department introduces not only its practices but also its philosophy, values, and standards to the employee. A progressive department will take a positive approach to this opportunity.

What does a positive approach mean? It means operating from the belief that recruit training exists to teach new firefighters what they need to know, in an environment that supports learning. Under such a philosophy, there is no room for a punitive approach that tries to make training as artificially difficult as possible in order to see who can "cut it" and who can't. Fire departments are finding it much more effective to emphasize learning over intimidation. Not surprisingly, this yields better results with recruits of all kinds.

Some instructors and chiefs feel that because the fire service is a paramilitary organization, recruit training should be done in militaristic ways. In reality, the fire service's paramilitary tradition is a half-hearted one. Current management philosophy shows a trend away from parallels with the military and towards the idea of the fire service as a business. Even where the paramilitary tradition was strongest, the fire service often adopted the worst aspects of the military (unquestioning obedience to all orders, the automatic superiority of officers) and few of its best (accountability, dedication to providing high-quality service, and control of inappropriate behavior). If particular aspects of paramilitary culture are important--such as when one must take an order without discussing it, or how the chain of command works--those can be taught readily. Other, counterproductive aspects of that culture should be revised. As one fire captain said,

> Our recruit academy want through at least ten years of so-called instructors who ran it like a boot camp. As an officer, I don't want an "Army grunt." I want a professional, wits-about-you firefighter.[3]

An approach to training that treats recruits with respect can develop their analytical skills and encourage them to be problem-solvers, rather than simply asking them to memorize information and repeat it on demand.

In practice, creating a positive learning environment means many things. For example, it means giving students plenty of time to practice physical skills before being evaluated on them. A common criticism of recruit training is that it does not provide this opportunity:

We had too much standing around, not enough hands-on time.

Not uncommon to do an evolution a couple of times and that's it.[4]

Instructors should give full, clear instruction on and demonstration of all physical skills. Students should practice the skill under supervision (for reasons of safety) and get instructor feedback to help them do it better. Only after each recruit has had the chance to go through this process should they be evaluated. Once the recruits have demonstrated their competence at each basic task, the tasks can be integrated into evolutions, and practiced and performed the way they would be on the fireground.

For evaluation, each recruit should complete each task for which they are responsible. Time constraints sometimes encourage instructors to "check off" students who in fact have only watched someone else perform a task. This is unfair to everyone concerned. It sets the recruit up to fail later on, in the field, and the failure will appear to be the recruit's fault: after all, he or she was "trained" and checked off on the skill.

A recruit training program can provide time for each student to learn, practice, and be evaluated fairly only if it receives adequate support from the department's leadership. If management does not make training a priority, the best-qualified and best-intentioned instructors will not be able to make it work. The recruit academy must be of a sufficient length to allow proper training, and either by limiting the size of the recruit classes or increasing the training staff, the ratio of students to instructors must be kept small.

A positive training environment also means having the flexibility to teach a wide range of recruits effectively. This relates to learning styles and backgrounds, as mentioned above, as well as to the ability to provide extra assistance to those who request it. This help may come from the training staff after hours, or from an official or informal mentoring program that links incumbent firefighters with recruits to provide support, guidance and supplemental training. If the training staff is to provide such help, their schedules must be designed to permit it: the program will work only on paper if instructors are so busy with their other duties that they never really have the time to help. However this assistance is provided, no penalties or censure should be attached to recruits' using it.

Technique

Most fireground skills and evolutions can be performed safely and efficiently by more than one method. Traditional approaches to firefighter training do not explore or even permit alternate methods; officers and instructors tend to view the traditional methods as "right" and then find a pretext for justifying them. One firefighter said

> I didn't realize how artificial this rigidity in techniques was until I got on my second fire department. In my first recruit school, I'd been taught that you always climb a ladder moving your right hand when your right foot moved, and your left hand with your left foot. They explained the reason for this, which I've forgotten by now. On my new department, we were doing ladder drills one morning, and the captain said, "Of course, you always climb right hand/left foot." And he had a reason for it, too. That started me wondering about all the other "absolutes" I'd been taught along the way, and how many of them were just as arbitrary?[5]

Firefighters should be encouraged to perform a task in the way that is most efficient for them and gets the job done safely. Physical techniques for smaller, shorter firefighters should be incorporated wherever possible into physical skills and evolutions. Different techniques should be demonstrated equally, without implying that one method is somehow inferior or should be used only if the recruit is having trouble with the "normal" or "regular" method.

For example, many women firefighters have found that using a reverse grip on a ladder halyard (with the thumb edge of the hands facing downward) allows a more efficient use of strength. Hose-handling techniques that allow a single firefighter to control an 1-1/2" or 1-3/4" line by crossing it in front of the body and allowing the firefighter's body weight to lean into it often are more effective for lighter-weight firefighters. Many such techniques can be learned by watching smaller veteran firefighters in action, or talking with women firefighters about the ways they accomplish fireground tasks.

Techniques that are more efficient for smaller, lighter firefighters usually also end up being safer and more efficient for others. "If the big guys use the same technique as the smaller people, they distribute the work more throughout their body, fatigue more slowly, and have less potential for pulling muscles," the training chief for one department observed. When evolutions require coordination of several people's efforts, some standardization of technique may be necessary for efficient operation, but efficiency, effectiveness, and safety should be the guiding criteria, not "how we've always done it."

Training instructors

Fire training instructors should be chosen for their ability to teach and communicate information to a wide range of students. They should be knowledgeable about firefighting, skilled at developing a curriculum and writing lesson plans, and interested in improving their department. They should have excellent organizational skills and a positive attitude toward training and toward the students. Instructors should never be assigned to the training academy as punishment, or because they need a light-duty assignment, or in order to improve their résumés.

The training staff should themselves be given appropriate education in how to teach, and should have frequent opportunities to attend training outside the department. They must have adequate time, resources, and departmental support to allow them to carry out their responsibilities. The training center should provide usable office space as well as classrooms, apparatus bays, a burn tower, and other necessary facilities. Equipment used in training should be contemporary with that used in the field: it is of little use to firefighters to be trained on out-of-date apparatus, and frustrating or dangerous to have to work with equipment that is cumbersome or malfunctioning due to age.

The training staff should be as diverse as possible. It should include women, people of color, and individuals of different ages and sizes. This is particularly crucial for classes that have women and minority recruits, but is an important general practice as well, to give the clear message that the fire department values the diversity of its workforce.

Consistency and fairness

It is crucial that information be presented in a consistent way to recruits: it is confusing and frustrating to the new firefighter to learn something from one instructor on Monday, only to have another instructor give conflicting information on Tuesday. Lesson plans should be in place for every lecture and hands-on training session, so the material will be presented consistently regardless of the instructor. (These lesson plans can also can be used by company officers in the stations, to train their crews.) Lesson plans should be reviewed on a regular basis to keep them consistent with department practice and equipment.

The criteria for passing or failing the fire academy should be explained clearly to all recruits and strictly adhered to by the instructors and staff. There should be no sources of "inside information" regarding tests, evaluations, or techniques; all recruits should have equal access to any useful knowledge. Evaluation procedures should be set up in advance, and appropriate forms designed to document each evaluation. To compensate for the subjectivity of evaluations, instructors should be trained in how to identify and reduce evaluator bias.

Ongoing training for incumbent firefighters

Have a comprehensive overall plan for your department's training. Input for this plan should come from all levels of the department. Since effective training for any occupation starts by assessing the desired outcomes, the department's managers should be asked to identify what they expect from their firefighters. A typical list might look something like this:

1. The basics of safe, effective, efficient, and consistent fireground and EMS operations.

2. "People" skills, both within the crew and with the public: the ability to interact productively with people in crisis, to listen and communicate effectively, to manage and help resolve interpersonal conflicts, and an awareness of diversity issues within the department and the community.

3. Knowledge of departmental policies, rules, and regulations, and the way things really work within the department.

4. The ability to interact efficiently with other agencies: police and other law enforcement departments, hospital emergency-room personnel, other city departments, other fire departments.

5. Health and fitness maintenance: exercise, diet, stress management, biomechanically efficient, and safe techniques for performing physical tasks.

Current training programs usually do not address all of these basic needs, and some of them are addressed only incompletely. A comparable list for officer training would reveal similar shortcomings in most departments' officer training programs.

Surveying firefighters about their training needs also provides valuable insight into the strengths and weaknesses of the department's training system. Survey questions should ask about training needs and priorities. To be most effective, the survey should be confidential, and department members should be able to see some results from it afterwards.

Company officers should be responsible for providing training for their crews, based on lesson plans and resources provided by the training staff. The training chief or manager must hold officers accountable for the quantity and quality of this training, and not allow training to happen on paper only. Multicompany training sessions should be held periodically. Individuals also should have the opportunity to train themselves by having access to training materials in the station or loaned from the training division. One firefighter suggested

> The training officer should be required to have lesson plans and job breakdown sheets for every drill, skill, and maneuver we do, organized in a binder for all stations as well as in the training office. All slides, videos and other training materials should be made available to personnel by providing a list and giving the officers the free use of the materials for their crews. There should be more class time to review drills, and the instructors should demonstrate the skills prior to the drill. Drills should be informative and creative rather than punitive.[6]

Conclusion

Good fire training creates a positive environment for new employees, improves the skills of current firefighters, and leads a fire department safely and progressively into the future. Bad fire training--or none at all--threatens the safety of all firefighters, reduces morale, particularly harms women firefighters' chances of success, and violates the department's prime directive to provide the best possible protection for the community it serves.

Notes:
[1] Floren, Terese M. "Survey Results: Firefighter Training." *Firework*, Vol. XIV, no. 6; April, 1996, p. 2.
[2] Brown, Cathy, "One Crazy Person Running In," *WFS Quarterly*, Vol. V, no. 4; Fall, 1990, p. 12.
[3] Floren, op. cit., p. 1.
[4] Ibid.
[5] Private communication to Women in the Fire Service, Inc.
[6] Floren, op. cit., p. 4.

Stopping sexual harassment in the fire service

Sexual harassment is...a form of sex discrimination. It is illegal, it can devastate those who experience it, and it often destroys the morale and productivity of the work environment. It is widespread in the fire service, and the numbers are not getting better. As many as 85 percent of women firefighters have experienced some form of sexual harassment at work or as volunteers.[1]

Sexual harassment is...a power play that is degrading, humiliating, and intimidating to its victim. It is based on aggression and hostility, not sexual desire. The physical appearance and behavior of the victim do not cause harassment. Sexual harassment is not "natural attraction," a "compliment" to the victim, or a "normal" way for men to react to women in the workplace. It is a way men assert their dominance over women and, in some cases, try to force women to leave the job.* Compliments that are welcomed, attention that is mutually desired, and truly friendly jokes and teasing do not leave either participant feeling uneasy or intimidated. Sexual harassment does.

Sexual harassment is...an intimate violation that often occurs without witnesses. Its victims generally feel powerless to stop it, which is what allows it to happen in the first place. Women in nontraditional jobs, often subjected to strong pressure to "go along to get along," and to fit in and be "one of the guys," are especially deprived of most forms of support in trying to stop workplace harassment. Women firefighters may come to believe that harassment is what they have to expect as a newcomer in a male-dominated workplace.

Sexual harassment is...anything from blatant acts such as physical assault and *quid pro quo* pressures ("Sleep with me, or you won't get hired") to more subtle behavior such as persistent, unwanted requests for dates, displays of pornography in the workplace, and jokes that put women in subordinate, sexual roles or call attention to their gender.

Sexual harassment is...a management problem for both legal and practical reasons. Employers are liable for acts of harassment that occur in the workplace, whether or not they themselves knew of the acts. Harassment makes the workplace hostile and unproductive for many in it. It can cause stress, poor job performance, heavy use of sick leave, and high employee turnover. Complaints and litigation result in time-consuming investigations and create poor public relations.

Why sexual harassment is not reported

Even thought it is illegal and a violation of most employer's policies, sexual harassment rarely is reported. Only about 5 percent of victims report incidents of harassment. Instead, a victim is much more likely to leave her job, request a transfer, or suffer in silence and hope the problem goes away. This is true in the fire service and is evidenced by comments from women firefighters who were harassed but chose not to report it:

> I didn't want to be labeled a troublemaker, and I didn't feel a positive outcome was possible.

> Much of the harassment occurred when I was on probation and felt I could not speak out...I did attempt to speak to my officers; all of them shrugged off my appeals for help.

> I didn't want to be "singled out" even more.

> I want to keep my job. It's clear that those who seek legal recourse can't come back to work.

> I didn't want to get someone suspended or fired. I just wanted it to stop.[2]

---

*The vast majority of harassers--95 percent--are men, and most victims of harassment are women.

41

There are many reasons women do not report sexual harassment. Some are complex and subtle, having to do with how women are brought up to view themselves and to behave. Others reflect the dynamics of a male-dominated workplace. Following are some of the most common reasons sexual harassment is not reported.

- Reporting the incident usually means an invasion of the victim's privacy. Harassment involves very personal interactions. Many women are uncomfortable with the prospect of having to discuss such subjects with their supervisors or other investigators, or of having to explain and relive the incidents. This is particularly true where the work environment in general is unsupportive, or where the victim has reason to believe the complaint process is not confidential. She may also fear, with good reason, that her own personal life or events in her past will be investigated, exposed, and presented as being somehow relevant to the harasser's behavior.

Verbal forms of harassment:

- sexual jokes or teasing;
- suggestive comments or sounds;
- remarks about a person's clothing, body, or sexual activities (real or imaginary);
- pressure or demands for dates or sexual activities;
- derogatory name-calling or sexual references; and
- threats or intimidation.

Visual forms of harassment:

- exposure to photographs, cartoons, drawings;
- gestures and other body language; and
- leering and other facial expressions.

Physical forms of harassment:

- touching, patting, squeezing, pinching, brushing against, cornering; and
- physical aggression or intimidation.

Survey of fire service women: sexual harassment

According to a 1995 Women in the Fire Service survey, 88 percent of women career firefighters and 83 percent of women volunteer firefighters had been subjected to one or more of the following types of harassment at some point during their work in the fire service. Most of the harassment was ongoing: 71 percent of career firefighters and 59 percent of volunteers were experiencing the behavior at the time of the survey.

- unwelcome requests or demands for sexual favors from coworkers or supervisors;

- unwelcome physical contact;

- sexually explicit or derogatory posters, photographs, or cartoons present or posted in the workplace that were offensive to the women;

- sexually explicit or derogatory movies or videotapes shown in the workplace that were offensive to the women; and/or

- sexual or sexist jokes or comments that were offensive to the woman.

Previously *quid pro quo* harassment was previously thought to be relatively uncommon in the fire service. Thus, it is startling to note that 25 percent of career women firefighters and 30 percent of volunteers responding to this survey had experienced unwelcome requests or demands for sexual favors in connection with their work or volunteer service.

- Victims of harassment often believe that pursuing the matter would do no good and might even make matters worse. Often these concerns are justified. Women on many fire departments have little reason to believe management will do anything to stop the problem. A woman may endure years of unwelcome, abusive behavior without ever finding herself in a position where filing a complaint would actually solve the problem.

- Victims also legitimately fear retaliation: that the harasser will increase the harassment, or that coworkers or the employer will strike back in some way. The more the victim depends on her income--for example, if she is single parent--and the fewer options she has for finding another job, the more likely she is to put up with harassment rather than jeopardize her paycheck by complaining.

- Because women are taught to be caretakers and nurturers, a woman may feel sorry for the harasser and not want to get him in trouble. She may try to find excuses for his behavior, or attempt to convince herself that something in her own behavior caused or contributed to the harassment.

- Victims of harassment may fear isolation and the loss of any friends or allies they have in the workplace if they rock the boat by complaining about the harassment. This is particularly likely in a workplace that fails to support those who report harassing behavior by filing a complaint.

A fire chief is, therefore, on shaky ground if he or she believes the department to be free of sexual harassment simply because no one has reported it. Ideally, an employee who is harassed will confront the harasser, report the behavior to a supervisor, or seek support from counselors or from friends who have been harassed. But most victims of harassment are disempowered and feel they have few options. This leads them to react in ways that neither stop the behavior nor let the employer know a problem exists: denial, trivializing or excusing the behavior, or trying to appease the harasser in the hope that he/she will stop. If a harassment victim takes any action, it is usually either to request a transfer or to quit his/her job.

What constitutes illegal harassment

Federal law defines illegal sexual harassment as:

Unwelcome sexual advances, requests for sexual favors, and other verbal or physical conduct of a sexual nature...when

(1) submission to such conduct is made either explicitly or implicitly a term or condition of an individual's employment,

(2) submission to or rejection of such conduct by an individual is used as the basis for employment decisions affecting such individual, or

(3) such conduct has the purpose or effect of unreasonably interfering with an individual's work performance or creating an intimidating, hostile, or offensive working environment.[3]

Helpful though it would be, it is impossible to make a comprehensive list of all behaviors that could constitute sexual harassment. Many situations must be examined individually; what is harassment in one case may not be in others. This can seem confusing until one focuses on the key concepts: "unwelcome," "intimidating or hostile," and "interfering with work performance." What is important is the **effect** of the questionable behavior on its victim. Behavior that is highly amusing to one person may be very unwelcome to another.

In cases that go to court, the court must determine whether the victim was offended by the behavior, and whether that reaction was "reasonable."

Courts have differing ways of determining what is "reasonable." Some use a "reasonable person" standard: besides being actually offensive to the victim, the harasser's conduct must be such that it would affect the work performance and psychological well-being of a reasonable individual. But because a woman's perspective may differ substantially from a man's, some courts have adopted a "reasonable woman" standard instead. (See "*Sexual harassment: the legal background,* " p. 55)

A "reasonable person" standard does not consider the difference between women's and men's views of appropriate conduct. For example, in one study, 67 percent of men surveyed said they would be complimented if they were propositioned by a woman at work. When women were asked if they would take such a proposition from a man in the workplace as a compliment, only 17 percent said yes.[4]

The chief's role in stopping sexual harassment

Many fire chiefs find it difficult to learn of harassment going on in their department. Harassment, after all, rarely goes on in front of those who could take steps to stop it. Other chiefs may be reluctant to deal with sexual harassment at all, believing that writing policies or providing training will increase the number of harassment complaints. But a head-in-the-sand approach is counterproductive: it will only make things worse, and it ignores management's legal responsibility for providing a harassment-free workplace.

Informal complaint processes

One police department's EEO unit has set up a complaint process that uses both formal and informal procedures. Employees who have been harassed may file a formal complaint, resulting in a full investigation by the EEO unit, or they may contact the unit with an informal complaint. When the unit receives an informal complaint, it sends a memo to the officer of the crew or station involved, with details of the problem as it was explained and suggestions for possible resolution. The officer then meets with the people involved, reaches a resolution, and sends a memo back to the EEO unit. The unit keeps the memos on file. A second complaint, formal or informal, results in a full investigation.

Some fire departments are now experimenting with informal complaint processes and find they are successful in stopping harassment that otherwise would not have been reported until it had worsened.

Complaints of harassment may, indeed, increase slightly for a short time after antiharassment training is done. This is a good sign. It means employees have learned that they don't have to put up with harassment as a price of their job, and that management will support them in getting problems resolved--not blame them for being the whistleblower. Employees who are knowledgeable about harassment and confident in management's stance against it are much more likely to try to resolve harassment informally, before it becomes a major problem.

Fire service leaders should work aggressively to prevent sexual harassment in their departments, and to deal with it promptly and effectively if it occurs. The attitude from the top should be one of "zero tolerance" of inappropriate, harmful workplace behavior. Following are some basic steps for managers to take in this area.

1. Adopt written policies prohibiting sex discrimination and sexual harassment, and a workable, confidential, step-by-step procedure for the filing of complaints. Give every employee or volunteer a copy of the policies. Include a list of individuals and agencies to which one can go with complaints.

2. Provide training for all personnel--firefighters, officers, and support staff--on interpersonal issues, including sexual harassment. Antiwoman behavior is part of the dominant culture of many fire stations and should be addressed as such. (See "Supporting Workforce Diversity" section, page 91. It may be helpful to copy and distribute the material on pages 51-54 of this handbook, which are designed to support such training.)

3. Do not ignore harassment. To do so sends the message that you agree with the behavior or its underlying attitudes. Do not place all of the burden for reporting and correcting the problem on the harassed person or targeted group. Each stage of prejudiced behavior encourages the next; extreme behavior will develop when subtle behavior is condoned.

4. Support people who bring complaints of harassment to your attention. Handle complaints with the utmost confidentiality, and treat all parties with respect.

5. Encourage solutions that stop harassment at the lowest possible level. An open-door policy, a multiple-option complaint procedure, and training in interpersonal communications all can go a long way toward resolving problems long before the chief is faced with a polarized situation or an expensive lawsuit.

6. Investigate complaints promptly and make decisions on them in a timely manner. Do not let the process drag on unnecessarily, particularly in situations where the harassment or the complaint are common knowledge.

7. Prevent retaliation against people who file complaints, and against any witnesses who support their allegations.

8. If the charge is found to have merit, discipline the harasser. Don't solve the problem by transferring the complainant/victim to a new station unless this is what he or she wants.

9. Find out the reasons behind an employee's request for a change in status, whether it is an application for transfer or a notice of resignation. Use exit interviews (interviews with people who leave the job) as a reality check.

10. Be a role model. Your language and actions should set an example for the rest of the department. Be aware of your own prejudices and make sure your behavior does not reflect them.

Notes:
[1] Women in the Fire Service, Inc., unpublished survey data, 1995. Earlier surveys by Women in the Fire Service (1990), by Diane Sanchez of Sunset Associates (1991) and Rosell, et al. (1995) found from 58 percent to 75 percent of women firefighters had experienced sexual harassment on the job. [Rosell, Ellen; K. Miller and K. Barber. "Firefighting Women and Sexual Harassment," *Public Personnel Management*, Vol. 24 no. 3 (Fall 1995): pp. 339-350.]
[2] Women in the Fire Service survey, 1990.
[3] EEOC Guidelines, 29 C.F.R. §1604.11.
[4] Goleman, D. "Sexual Harassment: It's About Power, Not Lust" *New York Times*, October 22, 1991: C1, C12.

Summary checklist for managers

- Adopt written policies prohibiting sex discrimination and sexual harassment, and give every employee or volunteer a copy.

- Provide antiharassment training for all personnel.

- Do not ignore harassment or inappropriate behavior.

- Support people who file complaints about sexual harassment, and prevent retaliation against them.

- Encourage solutions that resolve problems at the lowest level.

- Discipline the harasser, not the victim.

- Set an example of the kind of behavior you require of others.

Designing a policy against sexual harassment

Following are two policies against sexual harassment in the workplace. The first is a generic, "model" antiharassment policy that may suggest ideas or provide language for the development of a policy for your fire department. The second is an example of an effective policy currently in use. No policy should be implemented in your department without considering your department's specific needs or seeking qualified legal advice.

The _____ Fire Department is committed to providing a harassment-free work environment for all persons, regardless of their race, sex, religion, color, age, handicap, national origin, or sexual orientation. Any employee who engages in such harassment, who permits employees under his/her supervision to engage in such harassment, or who retaliates or permits retaliation against an employee who reports such harassment, is guilty of misconduct and shall be subject to remedial action which may include the imposition of discipline up to and including discharge. Retaliation against witnesses or persons who participate in the investigation of harassment also is misconduct, and offenders may be subject to discipline up to and including discharge.

In the workplace, constitutionally protected speech does not include ethnic or sex-related slurs, unwelcome sexual advances, the display of derogatory graphic materials, verbal or physical conduct of a racial or sexual nature, or other forms of harassing conduct. This type of behavior is unprofessional, and produces an uncomfortable work environment. It will not be tolerated.

[Some departments write the EEOC's definition of sexual harassment into their policy. If State law or local ordinances provide a more comprehensive definition, it also may be included. This information may be expanded by providing specific examples of unacceptable behavior such as the following.]

Examples of unacceptable behavior

Unwelcome or unwanted sexual advances, including patting, pinching, brushing up against, hugging, cornering, kissing, fondling, or any other similar physical conduct considered unacceptable by another individual;

Requests or demands for sexual favors, including subtle or blatant expectations, pressures or requests for any type of sexual favor accompanied by an implied or stated promise of preferential treatment or negative consequence concerning one's employment status;

Verbal abuse or kidding that is sexually oriented and considered unacceptable by another individual, including commenting about an individual's body or appearance where such comments go beyond mere courtesy; telling "dirty jokes" that are clearly unwanted and considered offensive by others, or any tasteless, sexually-oriented comments, innuendoes, or actions that offend others;

Engaging in any type of sexually-oriented conduct that would unreasonably interfere with another's work performance, including extending unwanted sexual attention to someone that reduces personal productivity or time available to work at assigned tasks;

Creating a work environment that is intimidating, hostile or offensive because of unwelcome sexually-oriented conversations, suggestions, requests, demands, physical contacts or attention, or the display of sexually-oriented posters, photos, or calendars.

[The more "entry points" there are into the complaint process, or ways a complaint can be filed, such as with the supervisor, through the City's EEO or personnel officer, the department's human relations committee, etc., the more likely it is that targets of harassment will make use of the procedure.]

Complaint procedure

Individuals who experience sexual harassment should make it clear to the offending person that such behavior is offensive to them.

Upon the occurrence of an act of sexual harassment, or upon repetition of such acts, the victim should immediately report the incident to her or his supervisor, to the department's EEO counselor, or the Personnel Department's sexual harassment counselor. All employees are assured that they may make such reports without fear of retaliation or reprisal by the city, department management, or their supervisors.

The employee has the right to speak in private with the person to whom the sexual harassment complaint is made, or to have a witness to the occurrences present.

Each complaint of sexual harassment will be investigated fully and completely by the department's EEO counselor or by the Personnel Department's sexual harassment counselor. All investigations will be handled with discretion, sensitivity, and due concern for the dignity of those involved, and will be as thorough as necessary. Anyone who is alleged to have committed acts of sexual harassment will be contacted during the investigation and permitted to make a statement.

All persons named as potential witnesses by the employee will be contacted as required during the course of the investigation. Any employee who has observed the incident(s) of sexual harassment should cooperate in the investigation. All employees are assured that they may cooperate in such investigation without fear of retaliation or reprisal by the city, department management, or their supervisors.

Employees may expect a timely resolution of all complaints.

> [The policy should inform employees that even if they do not use the city's complaint process, they do not lose the protection of State and Federal law. Information about State FEP and Federal EFO complaint procedures should be made available to all employees.]

An employee may take advantage of the formal grievance procedure in resolving the complaint. Additionally, an employee may file a complaint with (or seek advice from) the city's EEO coordinator without going to a departmental superior. The employee may decline to use the city's procedures and file a complaint directly with the State Department of Human Relations or with the U.S. EEOC.

Information and advice about sexual harassment may be obtained by contacting the Personnel Department's sexual harassment counselor.

Supervisors' responsibilities

All supervisors and acting supervisors shall maintain a working environment that is free and secure from occupational hazards, including sexual harassment. Any intrusion into the work location of any element that can cause an undue interference with an employee's performance of assigned duties shall not be tolerated. Supervisors should demonstrate by their own conduct that they are committed to providing a work environment free of harassment. Supervisors shall at all times refrain from harassment and retaliation and should counsel and instruct subordinates in defining and preventing harassment.

Supervisors or acting supervisors shall deal immediately with any act of sexual harassment of which they become aware. If the supervisor or acting supervisor is unable to obtain a resolution or feels the incident is likely to go to the grievance stage, he/she shall notify the EEO Officer within 24 hours. Every precaution shall be taken to ensure confidentiality at this informal information-gathering stage. This shall be followed by a written report.

The supervisor or acting supervisor shall immediately investigate any complaint of sexual harassment, and move to have any such incident resolved. Supervisors will notify the fire chief any time an employee complains of harassment or of conduct that could constitute harassment.

When employees report harassment by citizens, supervisors shall use all appropriate means to stop the harassing conduct. The supervisor shall inform the department or division head promptly about the report and what was done to resolve it.

Confidentiality

All employees shall cooperate in investigating complaints of harassment. The nature of harassment violations, particularly those involving sexual harassment, requires a high degree of confidentiality and flexibility in approaches to investigation and resolution. All employees shall keep their communications in such an investigation confidential and shall disclose them only to city officials and employees who need the disclosure in order to perform their duties.

U.S. Forest Service policy on harassment on the fireline

The national policy of the U.S. Forest Service ensures a harassment-free workplace. In fire and aviation management, our workplace is the fireline as well as the office.

Staff sometimes mistakenly consider a remote location or the incident environment enough of a departure from the usual workplace to depart from acceptable workplace behavior. We in the fire community must correct this misconception wherever we find it.

HARASSMENT IN ANY FORM IS NOT ACCEPTABLE AND WILL NOT BE TOLERATED. Each fire incident is an opportunity to reinforce the message of the Department of Agriculture and the Forest Service:

HARASSMENT--IT COULD COST YOU YOUR JOB AND A WHOLE LOT MORE

POLICY: The Forest Service will not tolerate harassment based on race, national origin, religion, age, mental or physical disability, color, gender, or any other factor such as sexual orientation, marital status, union affiliation, veteran's status, or political affiliation that might be used to categorize or identify any employee.

The Forest Service strives for a harassment-free work environment where people treat one another with respect. Managers, supervisors and all employees, as well as our contractors, cooperators and volunteers have the primary responsibility for creating and sustaining this harassment-free environment (by example, by job supervision, by coaching, by training, by contract enforcement, and by other means). All employees, contractor personnel and visitors must take personal responsibility for maintaining conduct that is professional and supportive of this environment.

ACTION REQUIRED: Managers and supervisors must take immediate action to stop harassment, to protect the people targeted by harassers, and to take all reasonable steps to ensure that no further harassment or retaliation occurs. Employees who witness harassment should report it to the proper authority.

LOCATIONS COVERED: The Forest Service work environment covers any area where employees work or where work-related activities occur, including travel. This includes field sites, government buildings and other facilities such as fitness centers and campgrounds. Also included are vehicles or other conveyances used for travel.

WHAT HARASSMENT IS: Harassment is coercive or repeated, unsolicited and unwelcome verbal comments, gestures or physical contacts, and includes retaliation for confronting or reporting harassment. Examples of harassment include, but are not limited to, the following.

Physical conduct: Unwelcome touching, standing too close, inappropriate or threatening staring or glaring; obscene, threatening or offensive gestures.

Verbal or written conduct: Inappropriate references to body parts; derogatory or demeaning comments, jokes or personal questions; sexual innuendoes; offensive remarks about race, gender, religion, age, ethnicity, sexual orientation, political beliefs, martial status or disability; obscene letters or telephone calls; catcalls, whistles or sexually suggestive sounds; loud, aggressive, inappropriate comments or other vocal abuse.

Visual or symbolic conduct: Display of nude pictures, (or pictures of) scantily clad or offensively clad people; display of intimidating or offensive religious, political or other symbols; display of offensive, threatening, demeaning or derogatory drawings, cartoons or other graphics; offensive t-shirts, coffee mugs, bumper stickers or other articles.

Individuals who believe they are being harassed or retaliated against should exercise any one or more of the following options as soon as possible:

• Tell the harasser to stop the offensive conduct, and/or

• Tell a manager or supervisor about the conduct, and/or

• Contact your Personnel Office, a Special Emphasis Program Manager, or any other individual you trust who would take action.

In addition, you may seek help from the Civil Rights Office, the Employee Dispute Resolution Office, your local Employee Assistance Program Office, or the R.O. Employee Relations Group.

PENALTIES: Any employee who engages in harassment will face consequences ranging from verbal warnings and letters of reprimand up to and including termination from employment, depending on the seriousness of the misconduct. Managers and supervisors who do not take action when they know or suspect that harassment is occurring will face the same range of consequences. Contractor staff who engage in harassment may be subject to comparable penalties from their employers, and a contractor who fails to enforce this policy may have its contract terminated. Visitors who harass may be removed from the workplace and prevented from returning.

Procedures for dealing with sexual harassment and handling harassment complaints

Your antiharassment policy also should explain in detail your fire department's or other agency's procedure for handling complaints of harassment. Following are some guidelines for setting up and discussing your department's harassment complaint procedure.

Encourage employees and volunteers to address harassment as soon as it occurs and at the lowest possible level. A typical first step is for the victim to inform the harasser that his behavior is unwelcome. This can be done in person (preferably in private, although the victim may prefer to have a witness present) or by letter. Informal, one-on-one efforts such as this help the victim gain a sense of control over the situation, give the harasser a chance to change the behavior, and often avoid a public confrontation and the need to file formal charges.

Another informal procedure fire departments may set up to deal with sexual harassment is mediation. Many women who have been harassed have said, "I didn't want to get anyone in trouble. I just wanted it to stop." Women who will not pursue formal complaints of harassment for fear of retaliation and isolation, and who are uncomfortable dealing one-on-one with their harasser, often are willing to take the matter to mediation. Mediation focuses on education and behavior, not punishment and blame. Like one-on-one efforts, it gives the victim some control over the outcome, preserves confidentiality and avoids a public confrontation. The presence of the mediator may make it easier for the victim to confront the harasser and for the two to work out a mutually acceptable solution.

Employees should not be required to use informal procedures before filing a formal complaint. Serious or outrageous cases of harassment rarely will be amenable to resolution at the informal level and are best dealt with through the complaint process. In addition, some victims may be unable to confront their harasser face to face, even with a mediator present. It should be up to the person who has been harassed to decide whether to use the informal procedures or to file a complaint.

The steps for filing a formal complaint should be spelled out clearly. The system should have multiple points of entry--for example, a victim might be able to file a complaint with her officer, with the harasser's officer (if this is not the same person), with someone who supervises both the victim and the harasser, or with an outside entity such as Human Relations or the EEO Officer. The options of filing a grievance through the union, or a charge with any city and State Fair Employment Practices agencies and the Federal EEOC, also should be spelled out.

The procedure should detail how the investigation will proceed. This should explain who will investigate: a mixed team of one man and one woman is often recommended, for example. It should reaffirm the commitment to a prompt and confidential investigation.

The procedure should include followup monitoring of the situation to make sure the harassment really stops and there is no retaliation by the harasser, coworkers, or supervisors. It also should include steps to review the victim's personnel records and remove any derogatory information that might have been added, particularly if the harasser was a supervisor or the harassment went on with the supervisor's knowledge.

Stopping sexual harassment: the station officer's role

All station officers or acting officers, whatever their rank, are responsible for maintaining a harassment-free work environment in their stations. Departmental procedures may assign the officer a specific role or responsibilities in the harassment complaint procedure, but even in departments that have no clear antiharassment policy, first-line supervisors are in a crucial position with respect to preventing sexual harassment.

- Stop any behavior occurring in your station or crew that is clearly harassment. In order to do this, you must be able to identify harassment when it occurs and to deal with it fairly, effectively, and quietly. If the training provided by your department on these issues is inadequate, use your own initiative to get the training and information you need.

- Educate your crew on what does and does not constitute harassment. This should be done as a part of departmental training, but if your fire department does not offer such training, or if the training has been inadequate, consider developing and implementing a comprehensive antiharassment training program for the people you supervise. Every fire department employee should have a working knowledge of sexual harassment as well as of the department's policies and procedures.

- If a member of your crew brings a complaint of sexual harassment to you, be supportive: they are alerting you to a problem in the workplace. Do not downplay the reported incident or attempt to make excuses for the behavior. Follow through on the complaint swiftly and impartially. Do not discuss the complaint with anyone not directly involved. Victims of harassment will not feel free to use the system if they fear the whole crew will know of their complaint immediately. The person accused of harassment also is entitled to confidentiality, particularly if the accusation is unfounded. Only those directly involved should know of the complaint; those called as witnesses should be cautioned not to discuss the matter with others.

- Prevent all forms of retaliation against the person who has filed the complaint. Harassment victims often believe that filing a complaint would do no good and probably would only make matters worse. If this kind of attitude prevails in your fire station, harassment will not be reported and will only escalate, creating a nightmare for the officer as well as for the victim.

- Do not overreact to harassment, but do not fail to act. Discipline should follow the department's established guidelines, and should reflect whether the offense is the individual's first or just the latest in a long series; it also must be appropriate for the severity of the behavior. Logically, a firefighter guilty of attempted sexual assault would be disciplined more severely than one who brought a copy of a pornographic magazine to work.

- Creating a harassment-free work environment includes stopping harassment by those who are not in the workforce. Be alert to discriminatory or harassing behavior by visitors to the stations: members of the public, friends of firefighters, equipment salespeople or repairers, and others. Handling these situations will require tact and diplomacy, but any person who behaves in that way must be made aware that you do not support their behavior or attitude, and that they will not be welcome in the station if the behavior continues.

If a coworker is being sexually harassed

Many workers who would never harass anyone themselves are guilty of tolerating harassment that occurs in their workplace. Some onlookers may mistake the victim's silence for acceptance; others may not want to get involved or may fear losing the camaraderie of their coworkers by being a "spoilsport." Fair-minded and supportive coworkers, however, will refuse to condone harassment in the workplace. Your support of the victim, as a friend or witness, can be crucial to her or him becoming aware that the behavior need not be tolerated.

- If you witness behavior at work that seems questionable, don't assume it's welcome just because the recipient doesn't complain. Talk with her privately, letting her know that you felt the behavior might be inappropriate. Especially if she is a new recruit or brand new in the station, remind her that she doesn't have to accept unwelcome sexual behavior, whether it involves comments, physical contact, pornographic posters, or videotapes. Show her that she has your support in getting the behavior to stop.

- If the behavior is clearly harassment but the victim is unwilling to take steps to stop it, talk with the harasser in private. Let him know that you find the behavior inappropriate and that it's not okay with you. If your own work environment is being made uncomfortable by the sexual behavior of others, you have the right to ask that the behavior stop, or to file a sexual harassment complaint yourself.

- If you think your own behavior towards a coworker, present or past, might be or might have been harassment, **stop the behavior.** To determine where to draw the line between friendly behavior or joking and harassment, consider:

  - Is there an equal and comfortable exchange?

  - Would you be doing the same thing to your mother or father? To someone of your own sex? If a member of your family were in the same room?

  - Would you want to see coverage of your behavior on the six o'clock news?

Ask the coworker about it privately. Let him or her know that the behavior will not continue if it's a problem, and that you can still be friends. If, after you've asked about the behavior, you're still not sure about it, stop the behavior anyway. There are hundreds of different ways for women and men to interact enjoyably and professionally in the workplace. Don't risk harming a positive work relationship by behaving in ways that make either of you uncomfortable.

If you are being sexually harassed

Fire department employees or volunteers who are the victims of sexual harassment should

- Know the department's policies prohibiting sex discrimination and sexual harassment, and the procedures for filing a complaint.

- Politely and firmly tell the harasser to stop, in front of witnesses if possible. If you are unable to confront the harasser directly, write a letter and give it to the harasser in the presence of a witness. Keep a copy of the letter in a safe place, not at work. Or speak to your supervisor, to someone from personnel, or to an EEO officer.

- If the harasser repeats the conduct, inform your supervisor immediately and follow it up with a note or letter to the supervisor. Again, keep a copy in a safe place, not at work. It is important that you give notice of this offensive behavior to your employer and that you have a record that you did so. (Courts generally will hold an employer liable for sexual harassment only where it can be shown that the employer knew, or should have known, about the conduct.)

- Document all incidents in a diary with time, place, names of witnesses, what was said or done, and an exact account of your response and any physical or emotional stress you experienced. Make sure your notes are accurate and focused on current and relevant concerns: if they are later used in a lawsuit, they might not be considered confidential. Keep this log in a safe place, not at work. Do not destroy your original notes: if you need to copy them over for clarity, do so, but be sure to keep the originals.

- If the behavior involved criminal conduct, such as rape, attempted rape, battery, or sexual assault, file a police report. Failure to report the crime may be used against you later on if you decide to file a lawsuit.

- Talk with other women on the job or who have left the job, especially those who have worked with the harasser in the past; they also may have been targets. Do not be surprised or discouraged if other women do not support your decision to fight or report the harassment.

- Keep records of all positive evaluations, promotions, etc., in the event that your complaint results in retaliation against you by your employer, officer, or other firefighters.

- Contact your union, labor organization, or employee group for assistance. If you work under a contract, be familiar with its antiharassment clauses, its provision for arbitration of grievances, and your rights under the bylaws of your union.

- Contact women's organizations for support. You are not the first woman to be victimized by this kind of behavior. Some chapters of the National Organization for Women (NOW) sponsor support groups for victims of harassment, and many chapters of 9 to 5 can give you support as well. Other women's professional and trade groups, such as Women in the Fire Service, may be able to offer advice and resources.

- Be aware that sexual harassment, and the decision to take action against it, are very stressful. If your employer offers an employee assistance program whose confidentiality you trust, use it. Try not to take the blame onto yourself. What happened to you was not your fault, the harasser did not "mean well," and he was not doing it because he likes you. You do not have to tolerate sexual harassment as the price of being a firefighter, and you do not have to adapt to the ways of your coworkers at all costs. If you have friends and coworkers who can offer support, depend on them. Be aware, however, that the general public is largely uneducated and often unsympathetic on this issue, particularly for women in nontraditional jobs.

- If informal or internal remedies fail to stop the harassment, if you lack confidence in the internal remedies (for example, if the person in charge of investigating your complaint is the one who is harassing you, or has threatened retaliation), or if you are retaliated against for complaining about harassment, you should consult an attorney. Look for a lawyer with experience in handling employment discrimination complaints. Try to obtain private counsel, even though local, State, or Federal human rights agencies may assign a staff attorney to handle your formal complaint. The women's bar association or working women's advocacy groups may be able to refer you to an experienced attorney.

- File a formal complaint with the Federal EEOC or with your State or local Fair Employment Practices (FEP) agency. There are specific time limits for filing such complaints, depending on whether the harassment is continuing or retaliation occurred, and whether your State or locality has an FEP agency. Title VII requires that you file with the EEOC within 180 days of the last act of discrimination. In areas with State or local FEP agencies, that time period is extended by 90 days if you filed with the local agency first; some States require you to file with the local agency first. To protect yourself, file as soon as possible. You may wish to file the charge and then ask that it be held while you attempt to resolve the problem internally. You can always drop the charge if matters are resolved to your satisfaction.

Sexual harassment: the legal background

In recent years, the sexual harassment of women workers has been in the headlines. The Clarence Thomas Supreme Court confirmation hearings and the revelations of sexual harassment in the military generated widespread outrage. In 1994, a San Francisco jury awarded plaintiff Rena Weeks, a legal secretary, $7.1 million in damages following her sexual harassment complaint against the world's largest law firm, Baker and McKenzie. (That substantial award was subsequently reduced by the judge and is the subject of appeal.)

Have attitudes and behavior changed much as a result of this attention to sexual harassment? Some thought the outcome of the Senate hearings--Thomas' confirmation to the Supreme Court--would discourage other women from asserting their rights. Instead, the EEOC saw a 71 percent increase in the number of sexual harassment complaints filed in the first 3 months after the hearings, and a 15 percent overall increase for the year.

The hearings also provided the impetus for Congress to pass the 1991 Civil Rights Act, which permits compensatory and punitive damages for victims of sex discrimination, not just an injunction and attorney's fees, as was previously the case. In the courts, a unanimous Supreme Court ruled in 1993 that an employee need not resign from her job to prove the harassment she experienced was serious and the conduct actually offended her.[1] (This case is discussed in greater detail below.)

In many workplaces, gender-biased behavior seems to be shifting from overt to more subtle forms of discrimination. And, although Federal law has dominated the field of discrimination since the enactment of Title VII in 1964, some State and local laws have exceeded Federal statutes to expand protection against discrimination based on sexual orientation, marital status, weight/obesity, etc., and other traits.

Defining sexual harassment

Sex discrimination on the job occurs when employment decisions or conditions of employment are based on an employee's sex instead of his or her qualifications. Sexual harassment is a form of sex discrimination that consists of unwelcome behavior toward an employee, by the employer or someone under the employer's control, that would not occur but for the person's sex.[2]

Not every type of inappropriate, even offensive, behavior in the workplace falls under the legal definition of harassment. For sexual harassment to be illegal under Title VII, the conduct must be **unwelcome** and must affect **a term or condition of employment.** The courts recognize two types of sexual harassment:

- **Quid pro quo** (Literally, "something given in exchange for something else.")

  This type of sexual harassment involves a loss of economic benefits to the employee who rejects her employer's sexual demands. In a case decided in January 1995, the Federal court for the Southern District of New York held that *quid pro quo* harassment is not limited to explicit sexual overtures: the crucial point is the exchange of job benefits for the toleration of sexual harassment.[3]

> ## Federal law defines illegal sexual harassment as
>
> Unwelcome sexual advances, requests for sexual favors, and other verbal or physical conduct or sexual nature...when
>
> (1) submission to such conduct is made either explicitly or implicitly a term or condition of an individual's employment,
>
> (2) submission to or rejection of such conduct by an individual is used as the basis for employment decisions affecting such individual, or
>
> (3) such conduct has the purpose or effect of unreasonably interfering with an individual's work performance or creating an intimidating, hostile, or offensive working environment.
>
> Source: EEOC Guidelines, 29 C.F.R. §1604.11.

- **Hostile environment**

  This type of sexual harassment occurs when abusive treatment motivated by the employee's gender creates an offensive or intimidating work environment. In determining whether a workplace has been "polluted" by sexual harassment and turned into a "hostile" environment, courts and other fact-finders (such as administrative EEO agencies) will look at the pervasiveness, frequency, length of time, and seriousness of the actions. A few isolated leers or rude comments probably will not constitute actionable harassment. On the other hand, no court would expect an employee to put up with serious physical assaults.

The courts vary widely as to what women employees are expected to tolerate. At one extreme is the conservative Sixth Circuit in *Rabidue* (a case decided in the 1980's) which took the position that for women in nontraditional employment, "what you see is what you get" and that any woman entering a traditionally all-male environment had better get used to the highly sexualized workplace because, after all, society condones the selling of *Playboy* and the open display of semi-nude pictures at the local newsstand and the broadcast of pornography on television.[4]

But in another 1991 construction workplace case, the court in *Robinson v. Jacksonville Shipyards* rejected the argument of the Sixth Circuit in *Rabidue*, saying the "social context" argument cannot be squared with Title VII's promise to open the workplace to women.[5] In a 1994 Eighth Circuit case, *Stacks v. Southwestern Bell*,[6] the court held that the employer could never have a legitimate reason for creating a hostile work environment.

Constant remarks about the physical attributes of women and the use of derogatory words or sexual terms to refer to women may also create a hostile work environment. Workplaces where this occurs are degrading and offensive to many women workers even if the women are not personally the object of the commentary.

Whether unintentional harassment or deliberate "hate speech," this kind of behavior serves to isolate women from their male coworkers.

Less clear-cut are calendars and other posted material that show women partially dressed and posed in sexually attractive ways ranging from "cute" to "seductive." In the fire service, some of the women shown are employed or portrayed as firefighters. These pictures suggest that a woman's appearance is somehow linked to her job performance. Such material can create confusion about sexual harassment, since many people only vaguely understand what does or does not constitute harassment. Material that encourages people to judge women firefighters on the basis of physical attractiveness rather than work performance is an insult to everyone's professionalism. All professional organizations, including the fire service, should reject degrading behavior that allows one group of people to be exploited in order to appeal to another group.

Hostile-environment harassment is particularly a problem for women who enter fields such as firefighting where rough language, sexual jokes, "centerfold" postings, sex-based videotapes or television programming such as X-rated cable channels, touching, gesturing, and frequent comments about clothing, body, lifestyle, and behavior are excused as "normal" behavior or treatment as "one of the boys."[7] Some courts, in dealing with workplaces "permeated by an extensive amount of lewd and vulgar conversation and conduct,"[8] have found in favor of employers if the person complaining initiated and participated in the conduct. This perspective, however, does not take into account the social pressures on women to become accepted by adapting to the sexualized male cultural norms of most blue-collar workplaces.

Glossary

EEOC: Equal Employment Opportunities Commission; the Federal agency in charge of handling anti-discrimination complaints under Title VII.

Plaintiff: The person or group of persons bringing a complaint in a court case.

Title VII: Title VII of the Civil Rights Act of 1964; the Federal law banning housing and employment discrimination on the basis of race, religion, sex, and national origin.

Examples of "quid pro quo" harassment:

- "If you'll go out with me tomorrow night, I'll make sure you get all the training you need to make your probation."

- "You're not driving the engine unless you give me a kiss."

- "I'm on the interview board for your promotion. If you expect me to give you a good rating, you'd better sleep with me."

Examples of "hostile environment" harassment:

- unwanted touches, kisses, and other personal/sexual attention;

- referring to women in derogatory, vulgar, and/or sexual terms; and

- posting of graphic or violent pornography.

"Nonsexual" sexual harassment

One still-developing area of sexual harassment law involves cases of "nonsexual" harassment where the conduct has no sexual content, but the only reason the victim is targeted is because she is female. While the EEOC definition of harassment refers specifically to conduct "of a sexual nature," courts have held that actions that contain no sexual element but are directed against an employee because of her or his gender are also illegal sexual harassment.[9]

Some fire department managers have tried to excuse illegal discriminatory conduct based on gender, even when it interferes with women's working conditions, as an acceptable or "natural" reaction by men to the entry of women into the workplace. In the fire service, such illegal conduct has included refusal to trade shifts, train, work with, or eat with the firefighter, as well as interference with her gear or personal property. In such situations, courts will look at the "pervasiveness" of the conduct (defined not only by length of time, but also by the offensiveness of the individual actions) and whether explicitly sexual harassment also took place.[10] A victim need not subject herself to an extended period of demeaning and degrading provocation before being entitled to seek the remedies provided under Title VII.[11]

Lewd language, pornographic materials, vandalism to personal property, and anonymous telephone calls can indicate a hostile work environment and be actionable under Title VII even if it is unclear who is responsible for the various acts.[12] Harassing conduct that has not yet caused psychological damage but has affected an employee's work habits and may seriously affect the psychological well-being of the employee in the future, may create an illegally abusive work environment.[13] Courts also have held that a hostile work environment may be created when employees are forced to observe others being sexually harassed: both the person being harassed and the observers might have a claim under Title VII.[14]

Unwelcomeness

Both the EEOC (as stated in its guidelines on sexual harassment) and the courts have made it clear that in order to decide whether sexual harassment has occurred, the fact-finder must determine whether the conduct in question was "unwelcome." In making that determination, the fact-finder--a judge, jury, or hearing officer--will look to the "record as a whole" and the context of the reported actions. Judging this record gives fact-finders considerable leeway.

The fact-finder first will attempt to determine **what occurred.** This requires a factual determination of the credibility of the witnesses. Often in these cases, there are only two witnesses: the plaintiff (the employee), and the defendant (the employer). Once questionable conduct is established, the fact-finder will have to decide **whether the plaintiff "welcomed" or "invited" the conduct.** The Supreme Court has said that fact-finders may allow evidence of a plaintiff's sexually provocative speech or dress in determining whether the sexual advances were unwelcome.[15] Rule 412 of the Federal Rules of Evidence, however, was amended in 1994 to expand the protection afforded alleged victims of sexual misconduct, and to encourage victims of sexual misconduct to participate in legal proceedings against alleged offenders. In a case involving alleged sexual misconduct (specifically including a sexual harassment action), the rule prohibits the admission of evidence that the victim engaged in other sexual behavior or of the victim's "sexual predisposition." Although there are certain exceptions to these limitations, the rule should make it more difficult for a victim's reputation as a "loose woman" to be used as proof that she "invited" or "welcomed" the harassment. Under the new Rule 412, evidence of an alleged victim's reputation is admissible only if it has been placed in controversy by the alleged victim.[16]

Several court decisions refused to find that the plaintiff's life excused the harassment she encountered at work. The Eighth Circuit held in 1994 that the lower court erred, at trial, in focusing on the plaintiff's private life, including the fact that she was dating a married man.[17] A year earlier, the same circuit held that a plaintiff's private life, no matter how reprehensible (plaintiff's nude photograph appeared in two different national motorcycle magazines), did not constitute acquiescence to unwanted advances at work by her employer.[18] A district court in Arkansas, however, held that the fact that a plaintiff kept vulgar cartoons and was able to repeat at trial very crude and lewd remarks made to her, eroded her credibility on the issue of whether those actions were "unwelcome."[19]

As a general rule, in making a determination of "unwelcomeness," courts look for clear and unequivocal expressions that the advances or other conduct of the alleged harasser are distasteful and unwanted. If a plaintiff sends mixed signals, the court may find the alleged harassing conduct was welcomed.[20] Similarly, the plaintiff who was previously in a consensual relationship with a person she claims is now sexually harassing her may have a greater duty to signal that the sexual conduct and attention that were once part of the relationship are no longer welcome.[21]

After establishing that the conduct under investigation was unwelcome, the fact-finder will attempt to determine **if the plaintiff was, in fact, offended.** Thus, a plaintiff who proves that conduct was unwelcome also must show that the conduct created a hostile environment **for her.** In judging a plaintiff's credibility on this issue, fact-finders often consider the plaintiff's age, background, and personal characteristics. A teenager on her first job might not be expected to put up with as much as a more "seasoned" individual. The Seventh Circuit held in 1994 that a plaintiff's conduct at work did not constitute an invitation to be a target for crude and vulgar speech and actions.[22] Three years earlier, that same circuit had found that a plaintiff's receptiveness to sexually suggestive jokes and activities **was highly relevant** and served to defeat her Title VII claim that she had been sexually harassed. In the earlier case, the plaintiff had been a civilian jailer who was an active participant in "reciprocating in kind."[23] A plaintiff who sends mixed signals--who has engaged in sexually-tinged pranks or has told racy jokes--may encounter skepticism when she argues that more of the same oppressed or intimidated her.

What plaintiff must show to establish that the conduct offended her

Before the 1993 Supreme Court decision in *Harris v. Forklift Systems,*[24] the circuit courts were split as to what a plaintiff must show to establish that the conduct offended her. Clearly, if she offered expert medical testimony that she had to be hospitalized for depression, that would constitute some "degree of injury." But the line separating offensiveness from mere annoyance was not clear. Former waitresses, "capable, outspoken," and "well able to handle the situation," had their claims dismissed.[25] In *Harris*, however, the Supreme Court held that "Title VII comes into play before the harassing conduct leads to a nervous breakdown." But the Court also seemed to disagree with the EEOC and a Ninth Circuit approach that the plaintiff need not show any actual injury: "If the victim does not subjectively perceive the environment to be abusive, the conduct has not actually altered the conditions of the victim's employment, and there is no Title VII violation." So a tough-skinned waitress--or firefighter--might still have difficulty showing actual injury or offense.

Another point to remember is that when determining whether a particular harasser's behavior was unwelcome, courts have recognized that victims will respond differently to behavior from different people. What may be welcomed from one person may not be welcomed from another.

Reasonableness standard

Before the Supreme Court's decision in *Harris*, the lower courts were divided on the issue of whose point of view should determine whether a work environment was "hostile." Some courts judged "hostility" by examining only the plaintiff's reactions to the workplace conduct. Other courts divided over whether the harassment should be judged from the viewpoint of the **reasonable victim,** the **reasonable person,** or the **reasonable woman,** without regard to the effect on the plaintiff. The Supreme Court in *Harris* adopted a two-pronged approach that determines both whether the plaintiff was offended by the workplace conduct and whether a reasonable person likewise would be offended. The reasonable-person component protects employers from the charges of "hypersensitive" employees, although it does not eliminate the impact of the social beliefs of the particular fact-finder applying it. Some courts, recognizing the differing perspectives between men and women as groups, continue to apply a "reasonable woman" standard when evaluating a hostile or offensive work environment.[26]

Length of time to file a claim

One of the issues plaintiff's advocates and lawyers face at the onset of a legal complaint is whether the victim is barred from filing a complaint because the applicable statute of limitations has expired. Victims of discrimination have a time limit in which they may complain. The length of time varies, depending on local statutes and where the complaint is filed. At a minimum, a charge under Title VII must be filed within 180 days of the alleged act of discrimination. In some States, a plaintiff may have up to 300 days.

Also at issue is how far back the victim can go to complain about previous incidents. The theory of **continuing wrong** allows victims to include a series of past events in their complaint and thus "toll" the statute of limitations (lengthen the time period in which a complaint can be brought). If a victim can prove a series of related acts and show the employer maintained a discriminatory system, the statute of limitations begins to run as of the last act committed in the series; thus, past acts can be included in the complaint. Defendants, for their part, will argue for dismissal under a kind of **"laches"** theory: that the plaintiff delayed too long in raising her complaints, to the detriment of the defendant. A court then might limit how far back the complaint could reach, although it is harder to sympathize with a defendant who is a long-time harasser or who failed to correct a long-standing problem.[27]

Liability

An employer may be held absolutely liable for acts of *quid pro quo* harassment by employees who function as supervisors or as surrogates for the employer. ("Absolutely liable" means the employer is liable even if it had no actual knowledge of the acts.) The courts are divided on whether absolute liability applies to "hostile environment" harassment. The Supreme Court has rejected blanket employer liability, but also has held that absence of notice to an employer does not always insulate the employer from liability, even when the employer established an antidiscrimination policy and grievance procedure, and the employee failed to use it. The Court also held that the employer's lack of actual knowledge will not necessarily protect the employer from liability. If the harassment is sufficiently pervasive, the employer will be assumed to have "constructive" knowledge.[28]

The EEOC's guidelines emphasize that employers have an affirmative obligation to prevent sexual harassment:

> Prevention is the best tool for the elimination of sexual harassment. An employer should take all steps necessary to prevent sexual harassment from occurring, such as affirmatively raising the subject, expressing strong disapproval, developing appropriate sanctions, informing employees of their right to raise and how to raise the issue of harassment under Title VII, and developing methods to sensitize all concerned.[29]

The EEOC's guidelines also state that an employer is responsible for harassing acts of its agents, regardless of whether the acts were forbidden by or known to the employer.[30] If an employer fails to investigate, conducts an inadequate investigation, does not follow its own grievance procedures, or fails to take appropriate corrective action, a fact-finder may find grounds for employer liability. Some cases have mentioned adopting a "reasonable employer" approach, comparable to the "reasonable person" standard for plaintiffs. Employers have also been held liable for the acts of nonsupervisory employees and even of nonemployees, where the employer knew, or should have known, of the conduct and failed to take immediate and appropriate corrective action.[31]

In addition to the employer being held liable for acts of sexual harassment in the workplace, the harasser may be held **personally liable** for such acts. That means that the harasser may be held financially accountable for any damages that occurred. In addition, depending on the jurisdiction in which the complaint is brought, the harasser's supervisor may be held personally liable.

Class-action lawsuits for sexual harassment

In 1991, a Federal judge approved the first class-action lawsuit for sexual harassment.[32] Previously, sexual harassment suits were brought only by individuals. Class action lawsuits have greater impact and permit harassment victims to share the costs of litigation (emotional as well as financial). Class actions also allow one plaintiff's testimony to support that of another. The *Eveleth* case was brought by three women iron miners claiming that all 100 women who worked or applied for work at the mine were subjected to discrimination in hiring and employment conditions, and to a pattern of sexual harassment and a hostile work environment.

Other remedies

Acts of sexual harassment also have been the basis for civil tort claims of assault and battery, or of intentional infliction of emotional distress. Other possible civil claims could include invasion of privacy, loss of consortium, breach of contract, and wrongful discharge. In cases of State and local government employees, it may be possible to bring an action under §1983 (42 U.S.C. §1983) to assert violations of due process or equal protection rights as a result of sex discrimination or sexual harassment.

Notes:

[1] *Harris v. Forklift Systems,* 114 S. Ct. 367, 126 L. Ed. 2d 295, 63 F.E.P. 225 (1993).

[2] The 1984 EEOC "Policy Statement on Sexual Harassment" defines sexual harassment as: "Unwelcome sexual advances, requests for sexual favors, and other verbal or physical conduct of a sexual nature…" (29 C.F.R. §1604.11).

[3] *Kariban v. Columbia University,* 14 F.3d 773 (2nd Cir. 1994).

[4] *Rabidue v. Osceola Refining Co.,* 805 F.2d 611 (6th Cir. 1986), cert. denied, 107 S. Ct. 1983 (1987).

[5] *Robinson v. Jacksonville Shipyards,* 54 FEP 83 (M.D. Fla. 1988).

[6] 65 FEP Cases 341.

[7] *DeAngelis v. El Paso MPOA,* 51 F2d 591 (5th Circ. 1995), *cert. petit.* 95-214, 64 LW 3172. Derogatory references to woman police officer in police newsletter did **not** constitute hostile environment. The court stated, "Title VII cannot remedy every tasteless joke or groundless rumor that confronts women in the workplace." Significantly, the plaintiff continued to receive good performance ratings, and the police chief publicly condemned the offensive columns. But see *Black v. City of Auburn,* 857 FSupp 1540 (M.D. Ala. 1994). Public use of "bitch" and "whore" creates hostile work environment and is not protected by the First Amendment.

[8] *Gan v. Depro Circuit Systems,* 28 F.E.P. 639 (E.D. Mo. 1982). Also see *Swentek v. USAIR, Inc.* 830 F.2d 552, 557 (4th Cir. 1987) finding that the appropriate inquiry is whether the complainant welcomed the particular sexual antics complained of and not whether she "was the kind of person who could not be offended by such comments and therefore welcomed them generally."

[9] *Olmer v. Iowa Beef Processors,* 66 F.E.P. 843 (D. Neb. 1994). This is a particularly outrageous case of a continual and prolonged pattern of verbal and ultimately physical harassment by the supervisor and a coworker of the only woman on a 15-person maintenance crew. Although the harassment was not "sexual" in nature, it was clear to the court that it was aimed at the plaintiff solely because she was female, and the court awarded the plaintiff back pay and attorney fees. Also see *McKinney v. Dole,* 765 F.2d 1129, 1138 (D.C. Cir. 1985); *Hall v. Gus Construction Co.,* 842 F.2d 1010, 1013 (8th Cir. 1988); *Hicks v. Gates Rubber Co.,* 833 F.2d 1406, 1415 (10th. Cir. 1987); *Bell v. Crackin Good Bakers Inc.,* 777 F.2d 1497 (11th Cir. 1985). However, in *Walk v. Rubbermaid, Inc.,* 69 F.E.P. 1577 (N.D. Ohio 1994), the court found that the plaintiff's supervisor's statement that he had "no time for you or your f_____ menopausal bitches" was, standing alone, insufficient evidence to show that the supervisor's offensive behavior was gender-based, since both men and women were offended by the supervisor's remark.

[10] See *Carrero v. New York City Housing Auth.,* 890 F.2d 569, 578 (2d Cir. 1989).

[11] *Andrews v. City of Philadelphia*, 895 F.2d 1469, 1485 (3d Cir. 1990): "The pervasive use of derogatory and insulting terms" could be probative of a sexually hostile environment.

[12] *Id.*

[13] *Howard v. Dept. of Air Force*, 877 F.2d 952, 955 (Fed. Cir. 1989).

[14] *EOC v. Gurnee Inn Corp.*, 48 F.E.P. 871, 879 (N.D. Ill. 1988), *aff'd*, 914 F.2d 815 (7th Cir. 1990).

[15] *Meritor Savings Bank v. Vinson*, 477 U.S. 57 (1986).

[16] PFed. R. Evid. 412(a)(1), (2). As an interesting contrast, see *Heyne v. Caruso*, 69 F.2d 1475, 69 F.E.P. 408 (9th Cir. 1995). In that case, the Circuit Court of Appeals found that the trial court erred by refusing to admit evidence of the defendant's sexual harassment of other female workers, since such evidence was relevant to prove his motive or intent in discharging the plaintiff. The court held that the sexual harassment of others, if shown to have occurred, is relevant and probative of the harassing party's general attitude of disrespect toward female employees.

[17] *Stacks v. SW Bell*, 27 F.3d 1316.

[18] *Burns v. McGregor Elect.*, 989 F.2d 959 (1993).

[19] *Tindall v. Housing Auth.*, 762 FSupp. 259 [WD Ark. 1991].

[20] *Sardigal v. St. Louis Nat'l Stockyards Co.*, 42 F.E.P. 497 (S.D. Ill. 1986). Plaintiff alleged sexual harassment and sexually offensive environment based on fellow employee's conduct. Plaintiff was found to have "welcomed, if not encouraged" the sexual conduct and remarks based on plaintiff's voluntarily visiting employee, going for a drive with him and allowing him to visit her at her home. *Reichman v. Bureau of Affirmative Action*, 536 F. Supp. 1149, 30 F.E.P. 1644 (M.D. Pa. 1982). Plaintiff found not to have been subjected to "unwelcome" advances because she continued to invite supervisor to dinner and flirted with him after the incident complained of.

[21] *Koster v. Chase Manhattan Bank*, 687 F. Supp. 848, 46 F.E.P. 1436 (S.D. N.Y. 1988). See also, *Walker v. Sullair Corp.*, 736 F. Supp. 94, 52 F.E.P. 1313 (W.D.N.C. 1990) *aff'd in part and rev'd in part*, 946 F.2d 888 (4th Cir. 1991). Because there was no evidence that plaintiff had ever resisted the advances of her manager, and because the relationship began with the plaintiff's consent, the court could not find plaintiff had been subject to sexual harassment in her romantic affair with the manager.

[22] *Carr v. Allison Gas Turbine*, 32 F3d 1007.

[23] *Reed v. Shepard*, 939 F2d 484. However, if a person says "Stop!", that direct refusal could be probative of the "unwelcomeness" of the behavior, even if the person had previously "acquiesced."

[24] *Harris*, op. cit.

[25] *Kirkland v. Brinias*, 741 FSupp. 692 (ED Tenn. 1989) *aff'd* 944 F2d 905 (6th Cir. 1991).

[26] *Steiner v. Showboat Operating Co.*, 25 F3d 1459 (9th Cir. 1994).

[27] *Palmer v. Kelly*, 17 F3d 1491 (DC Cir. 1994); *Gomes v. Avco Corp.*, 964 F2d 130 (2d Cir. 1992).

[28] *Meritor Savings Bank v. Vinson*, op cit.

[29] 29 C.F.R. §1604.11(f).

[30] 29 C.F.R. §1604.11(c).

[31] Nonsupervisory coworker EEOC Guidelines 29 C.F.R. §1604.11(d); cases include *Hall v. Gus Construction*, 842 F.2d 1010 (8th Cir. 1988) [incidents of harassment so numerous that employer liable for failing to discover what was going on and remedy it]. According to case law, the non-supervisory harasser is not himself liable under Title VII [he may be under tort or other local law] because he is neither an employer nor an agent of an employer under Title VII. Nonemployee EEOC Guidelines 29 C.F.R. §1604.11(e); cases on the issue of employer liability for harassment by nonemployee(s) include *EEOC v. Sage Realty Corp.*, 507 F. Supp. 599 (S.D.N.Y. 1981) [revealing uniform led to many incidents of harassment by nonemployees].

[32] *Jenson v. Eveleth Taconite Co.*, 139 FRD 657 (D.C. Mn.).

The legal background: sexually explicit materials in the workplace

The ability of fire departments to regulate sexually explicit materials in the workplace seemed to have been dealt a blow by a U.S. District Court's 1995 decision in *Johnson v. County of Los Angeles Fire Department*,[1] which struck down that department's policy regarding *Playboy* magazine. That policy stated

> The following types of sexual material are prohibited in all work locations, including dormitories, restrooms and lockers...Sexually-oriented magazines, particularly those containing nude pictures, such as *Playboy, Penthouse*, and *Playgirl.*

A male fire captain (Johnson) sued the department, claiming the policy violated his First-Amendment right to possess, read, and provide "consensual sharing" of *Playboy*.

With the assistance of *Playboy* and the Los Angeles chapter of the American Civil Liberties Union (ACLU), Captain Johnson argued that the policy posed a particularly severe restriction on firefighters because the fire station operates as firefighters' *de facto* home for consecutive days and because "relaxation" is important to firefighters' mental health. He also claimed that the design of fire stations, with separate sleeping cubicles, provided enough privacy that he could "quietly enjoy" and "consensually share" *Playboy* without exposing it to unwilling viewers.

In striking down the policy with respect to *Playboy*, the judge did not consider the issues raised by more graphic magazines that were also banned. Instead, he focused on testimony from *Playboy* that the magazine itself is a paragon of equal opportunity with a woman president and many women readers, that the magazines does not depict sexual intercourse but only female nudity, and that most of the magazine is devoted to excellent articles on literature, art, politics, and so forth. The judge also attached considerable weight to the testimony of two women firefighters who said they were not offended by *Playboy* in the firehouse (as opposed to three women firefighters who were offended). Concluding that until discriminatory thought is "manifest," it is outside the scope of Title VII, the judge found that the defendant (the fire department) had not provided evidence establishing that quiet reading of *Playboy* contributes to a sexually harassing workplace.

In reading the decision, one might question the vigor with which the defendant fire department defended its policy. No evidence was offered about the effect of the presence of more graphic magazines in the workplace. No evidence was offered of the national scope of the problem of sexual harassment in the fire service, and how fire service women have suffered in their careers because of the polluted atmosphere in the firehouse that is fostered by the sexual objectification of women in magazines featuring female nudity. Indeed, the department did not even cite its own problems of sex discrimination and harassment, which were severe enough that within several months the ACLU itself filed a lawsuit against the county for pervasive sex and race discrimination.

The relevant consideration in determining whether firefighters should have access to *Playboy* 24 hours a day, 7 days a week, might have been the fire service's level of commitment to a harassment-free environment, rather than *Playboy* magazine's level of commitment to equal employment opportunity. The department also did not adequately address the problem of women firefighters as a captive audience who cannot escape the effects of the presence of magazines of this type by simply averting their eyes. The judge in the case upheld the right of men to read sexually explicit materials while being paid by taxpayers over the right of women to be free of sexual stereotyping in the workplace or of the fire department to regulate the presence of sexually explicit materials on its premises.

In contrast, in an earlier case in another jurisdiction, attorneys representing welder Lois Robinson, who had filed a complaint over sexually explicit materials in her workplace, presented the court with substantial evidence that those nude, seminude, sexually suggestive, and submissive pictures of women created a hostile environment for women workers. Using expert witnesses at trial, the plaintiff established that the pervasiveness of such materials led to sex-role stereotyping by her male coworkers, creating a discriminatory workplace for women. The court found evidence that pornography causes "behavior [that] is not directed at a particular individual or group of individuals, but is disproportionately more offensive or demeaning to one sex" and that such behavior

> creates a barrier to the progress of women in the workplace because it conveys the message that they do not belong, that they are welcome in the workplace only if they will subvert their identities to the sexual stereotypes prevalent in that environment.[2]

The *Robinson* court took particular note of evidence regarding the gender imbalance in the workplace and a woman's solo or nearly solo status as having concrete effects on both men's behavior and women's responses.

Notes:

[1] 66 FEP Cases 205 (C.D. Ca. 1995).

[2] *Robinson v. Jacksonville Shipyards*, 54 FEP 83 (M.D. Fla. 1988), at 1522-23.

Reproductive safety and family issues for firefighters

Firefighters, male and female, face hazards to their reproductive health in the course of their work. These hazards include such common fireground exposures as heat and carbon monoxide, as well as a wide array of less frequently encountered risks. Exposure to such hazards may make it more difficult to conceive a child or less likely the fetus will be carried to term. Exposure may affect fetal development and possibly cause birth defects. Toxins may be transmitted through the bloodstream to the fetus, or to the infant during breastfeeding.

Concern over the issue has arisen in a roundabout way. Not long after the first women became career firefighters, some fire chiefs voiced their awareness of the reproductive risks of firefighting. Typically, this was expressed less as a concern for worker safety than as an argument against women being firefighters at all. Rather than taking steps to address the risk, the fire service in general saw the ability to get pregnant as further evidence that women didn't belong on the job.

Women firefighters in those early years rarely had the option of planning a family. Many felt that, in being hired as a firefighter, they had tacitly agreed not to get pregnant. Women who did get pregnant sometimes faced the bleak prospect of losing their jobs, and were almost always in the uncomfortable position of not knowing how their pregnancy would affect their employment or their income.

As recently as 1986, only 10 percent of U.S. fire departments had maternity policies specifically designed for fire personnel.[1] Most policies that did exist focused only on pregnancy; male firefighters often remained completely unaware of the risks their job might pose to their reproductive potential and to their children. Only in the 1990's have fire service leaders begun to take an interest in safeguarding the reproductive health and safety of all personnel.

In addition to protection from risk, working parents of both sexes need time off work when their child is born, adopted, or seriously ill. The majority of employees will need this kind of leave at some time during their careers. Recent changes in U.S. law have set a minimum standard for family leave; fire departments may wish to implement policies more substantial than those required by law.

Risks to firefighter reproductive health

The research on firefighting and reproductive safety is too sparse to support solid conclusions; much more research is needed on firefighters' exposures to heat, noise, and toxic chemicals, and on the impact of these exposures on reproductive health. The information that is available, however, clearly indicates the presence of significant risks.

One of the most common products of combustion is carbon monoxide, to which the developing fetus is particularly vulnerable. Fetal hemoglobin has a higher affinity for carbon monoxide than does the mother's hemoglobin. This means the fetus will be affected more than the mother; exposures that produce only moderate symptoms in the mother can be fatal to the fetus.[2]

A father's employment as a firefighter may result in a higher risk of congenital heart defects for his children. A study in British Columbia in the late 1980's found a linkage between increased rates of atrial and ventricular septal defects in the children of male firefighters. The study concluded, "The occupation of fire fighting merits further attention in view of increasing knowledge regarding male-mediated teratogenesis (birth defects)."[3]

In 1989, researchers at Johns Hopkins University gathered reproductive data from firefighters. Their research found that chemical exposures pose risks to the reproductive health of both male and female firefighters, while carbon monoxide poses risks to the developing embryo.[4] High ambient heat in the firefighter's working environment also causes problems. The core body temperature of firefighters involved in interior operations at structure fires can rise high enough to impair sperm production in male firefighters and pose a risk of birth defects to an embryo carried by a pregnant firefighter. [Another medical authority has linked maternal hyperthermia early in pregnancy to neural tube defects in the fetus.][5] and maternal exposure to high levels of noise--firefighters' exposure include sirens, air horns, vehicle engines, and power tools--is linked to decreased fetal weight and increased fetal mortality.[6]

Preliminary results from a 1995 survey conducted by Women in the Fire Service showed an incidence of birth defects, premature births, and other childbirth problems in the children of women firefighters that may be higher than normal. The incidence was significantly higher when the child's father was also a firefighter.[7]

One group of EMS workers found high rates of gynecological problems in women EMT's and paramedics who worked with 800 mHz radios or in close proximity to video display terminals mounted in their ambulances. Menstrual irregularities, miscarriages, birth defects, uterine cysts, cervical abnormalities, and cancer were found in as many as 100 of the service's 680 women workers. According to the head of a group investigating the problem, when the terminals and radios were first installed, 15 women at one station alone developed menstrual irregularities.[8]

Glossary

Hemoglobin: the chemical compound in the blood that carries oxygen to the tissues

Hyperthermia: high body temperature

Teratogenetic: causing birth defects

For the nursing mother, the risk continues after the child is born. Studies on cigarette smoking and other chemical exposures make it appear likely that the same or similar chemicals in the fire environment are passed on to the infant through maternal milk.[9]

Fetal and maternal susceptibility to toxins in the workplace often have been used as an excuse to keep women out of particular jobs. Once it is shown that men, too, are harmed by these toxins (vinyl chloride and lead are two examples), the toxins are regarded in a new light: they become "workplace hazards." The prevailing view quickly alters to find ways to manage exposure risks rather than prevent employees from working. The same evolution of attitude and policy will probably hold true for the fire service. According to one of the Johns Hopkins researchers,

> For the few reproductive toxins that have been well studied, evidence demonstrates effects mediated through both males and females. In fact, one author has suggested that males may be more sensitive to exposure to reproductive toxins."[10]

The legal background

Two Federal laws and two court decisions provide the framework for developing reproductive safety and family leave policies in the workplace. The first law is the Pregnancy Discrimination Act of 1978 (PDA), an amendment to Title VII of the Civil Rights Act of 1964. The PDA broadens the definition of sex discrimination to include discrimination based on pregnancy and childbirth. It states

> Women affected by pregnancy, childbirth or related medical conditions shall be treated the same for all employment related purposes…as other persons not so affected but similar in their ability or inability to work…[11]

This law applies to all employers with 15 or more employees, and to employment agencies and labor organizations.

The PDA was the first significant piece of legislation to deal with maternity in the workplace. To prevent arbitrary and discriminatory treatment of pregnant employees, it guaranteed pregnant women access to benefits already in place for other workers. For example, company health insurance plans no longer could exclude coverage for pregnancy and childbirth. An employer could not refuse to hire or promote a woman solely on the basis of pregnancy. Disability caused by pregnancy had to be covered under an existing disability program. If the employer gave temporarily disabled workers light duty, it also had to give pregnant employees light duty. Most importantly, a woman could not be fired arbitrarily from her job if she became pregnant, nor could she be required to take an extended leave that was not medically necessary.[12] The PDA allowed women to continue their employment well into pregnancy and to return to work as soon as they were physically able.

In the years following the passage of the PDA, several States developed policies mandating certain types of benefits for pregnancy and childbirth. California passed a law that guaranteed up to 4 months of unpaid leave for the purpose of childbirth and recovery. The California Federal Savings and Loan Association challenged this law as a violation of the Pregnancy Discrimination Act, arguing that since the PDA required all employees to be treated the same, special benefits for pregnancy and childbirth were illegal.

The case went to the Supreme Court, and in 1987 the Court upheld California's policy and clarified the intent of the Pregnancy Discrimination Act. The majority opinion stated that the Federal law was intended to "construct a floor beneath which pregnancy benefits may not drop, not a ceiling above which they may not rise."[13] This decision allowed individual employers to develop policies specifically for maternity.

Some fire departments, as well as employers in other potentially hazardous work environments, developed policies for employee pregnancy based on the concept of "fetal protection." Chemical and environmental factors in many workplaces offer probable or proven hazards to a developing fetus and to the reproductive health of both men and women workers. Policies based on fetal protection often required a woman to leave a certain type of job, such as active firefighting, at some point in her pregnancy; in many cases, as soon as the pregnancy was known. This requirement was based not on the woman's ability or inability to perform her job, but rather on the employer's concern that some aspect of her work might prove harmful to the developing fetus.

The Supreme Court addressed fetal protection policies in its 1991 decision in *UAW v. Johnson Controls*. This case concerned a battery manufacturer that had barred all fertile women from working in jobs using lead because lead could harm a fetus. The Supreme Court rejected this policy as a form of sex discrimination and a violation of the Pregnancy Discrimination Act. The majority opinion stated: "Decisions about the welfare of future children must be left to the parents who conceive, bear, support, and raise them rather than the employers who hire those parents." It further pointed out:

It is no more appropriate for the courts than it is for individual employers to decide whether a woman's reproductive role is more important to herself and her family than her economic role. Congress has left this choice to the woman as hers to make.[14]

Although other decisions relating to pregnancy and employment exist in the lower courts, no Supreme Court decisions have modified or contradicted the thrust of *California Federal Savings & Loan* and *Johnson Controls*.

The second Federal law is the Family and Medical Leave Act (FMLA) of 1993, which guarantees 12 weeks per year of unpaid leave to employees in order to care for a newborn, newly adopted, or seriously ill child. The employer must continue health care benefits during this leave, and must reinstate the employee into his or her original position, or into an equivalent position with equivalent pay and benefits. Employers may require employees to use any paid sick leave and vacation time as part or all of the twelve weeks of FMLA leave. (*For more information on the Family and Medical Leave Act, see pp. 71-73.*)

Parental leave is not a new concept. The parental leave policies of six European countries (Austria, Germany, France, Italy, Finland, and Sweden) guarantee from 12 to 52 weeks of leave, with 60 to 100 percent salary retained during all or part of the leave. Canada offers 17 to 41 weeks of parental leave, with 15 weeks at 60 percent salary guaranteed. By contrast, the 12 weeks of leave mandated by the FMLA are entirely unpaid.

Developing policies for fire departments

Policies may be developed in a number of ways. It is more effective for fire department management to draft the policy after consulting legal counsel and employee representatives. City or county administrations also may draft such policies, but should be advised closely by fire department personnel in doing so. Where neither the fire department nor the city will implement a suitable policy, the matter often is negotiated by the firefighter's union. This somewhat distorts the purpose of the policy, as it converts workplace safety into a benefit and subjects it to the tradeoffs of the bargaining table. One advantage, however, is that male-dominated union locals are more likely to push for all-inclusive reproductive safety policies, rather than those affecting pregnancy only.

Although employment policies that relate to pregnancy and childbirth often are grouped under the general heading of "maternity" policies, three separate areas of risk or need should be addressed

1. Reducing reproductive risk for the pregnant employee, the breast-feeding employee, and the male or female employee attempting to conceive a child.

2. Providing adequate leave for the woman during the time she is disabled as a result of childbirth.

3. Providing adequate leave for new parents surrounding the birth or adoption of a child.

Reducing risks to reproductive safety. The general trend among career-level fire departments is to guarantee alternate, nonhazardous duty to a firefighter during the term of her pregnancy. All women workers need to be able to take leave from their jobs for the time surrounding childbirth, but women firefighters' needs are more complex than that. While the PDA guarantees pregnant women the right to stay on the job as long as they are able to perform, there is real concern as to whether women firefighters should continue in their usual job assignments while pregnant.

Many physicians agree that women should stop fighting fires and doing other high-risk work at some point in their pregnancies. Exactly what point is most appropriate is a subject of some debate. Because many environmental hazards are most dangerous to a fetus during the first trimester (the first three months a woman is pregnant), many fire departments want women to leave hazardous duty as soon as their pregnancies are known. The majority of women prefer this arrangement as well.[15] Nonetheless, the Supreme Court decision in *Johnson Controls* strongly suggests that such transfers cannot be mandatory.

As of 1996, the law upholds a woman's right to continue working in a hazardous environment even during pregnancy. If faced with a choice between a higher-risk pregnancy and the economic devastation caused by a significant loss of pay and benefits during pregnancy (because no alternate duty is offered), many women will feel compelled to choose the former. The International Association of Fire Fighters (IAFF) in 1992 adopted the position that the pregnant firefighter should be offered the opportunity for voluntary transfer from firefighting at any time during her pregnancy without loss of pay or benefits.

As relevant research increases, more attention is being given to the reproductive risks of parental exposure near the time of conception. Women and men who are trying to conceive should avoid environments that may endanger their reproductive health. Thus, nonhazardous duty should be offered to employees attempting to conceive a child. Because of the potential for toxins to be transmitted to an infant through breast milk, nursing mothers also should have the option of alternate duty.

Alternate, nonhazardous duty should be meaningful work that does not penalize the employee and should involve no loss of pay or benefits. Firefighters on alternate duty have worked productively in such areas as training, public education, prevention and inspection, policy development, and communications. Fire departments of all sizes have found ways to use these employees productively in roles that do not involve reproductive hazards; the department and the employee can both benefit from a short-term assignment to a noncombat position.

Nonhazardous assignments for workers seeking to avoid reproductive risks are not necessarily the same as "light" duty that may be made available to those returning from an injury. For example, light duty might require a firefighter to fuel staff cars every day, but excessive exposure to gasoline and diesel fumes during pregnancy may cause health problems. Similarly, doing arson investigation in freshly burned buildings is probably not the best choice of assignment for a firefighter who needs to avoid reproductive risks.

Firefighters on nonhazardous duty should be allowed to take training or recertification classes that other firefighters are undergoing, as long as the classes do not involve risks. This not only keeps them from falling behind in their training and saves the department from having to train them later, but also allows them to remain in touch with coworkers and with suppression operations while they are reassigned.

A fire department is not required to offer alternate-duty assignments to pregnant firefighters unless it has a policy of reassigning all employees who have temporary disabilities. Pregnant employees may, however, be offered alternate duty even if that option is not available to other employees. The legality of this type of variance in treatment was clearly upheld in *California Federal Savings & Loan*.

Maternity leave. Only a generation ago, motherhood and work outside the home were considered mutually exclusive. If a woman had a job at all, she was expected to quit once she started having children. Many employers refused to deal with maternity as an issue in the workplace. Employers were allowed to treat pregnant employees in any way they chose, and the law offered no protection for these women. As recently as 1964, 40 percent of all employers terminated workers who became pregnant.

Even after women had been in the career fire service for years, many of their employers had not developed policies to address pregnancy. A 1995 survey showed that 60 percent of fire departments that employed women had no written maternity policy, and an additional 23 percent had only a city- or county-wide policy that did not specifically address the needs of women in hazardous professions. Only 30 percent of fire departments employing women had a written maternity policy specifically for firefighters, and many of these simply consisted of language specifying that pregnant women could use sick leave and vacation time, or (as Federal law requires) that departmental provisions for employees injured off duty applied to pregnant firefighters.[16]

Maternity leave policies addressed the period of time when a woman is physically unable to work as a result of pregnancy and childbirth. All women who go through childbirth are temporarily disabled by it to some degree, whether for a few days or for several months. Every workplace should have a policy that allows women to have children and physically recover afterwards. When a woman's pregnancy does not restrict her ability to do her job, and when she suffers no complications from the pregnancy or delivery, a typical maternity leave might be 6 to 8 weeks. Extensions should be available for cases where the pregnancy or delivery are very difficult. Employers sometimes provide this time as paid or partly paid leave specifically for childbirth; in other instances, women must use sick or vacation time to maintain their income.

In a minority of cases, a pregnant employee will be unable to work even in an alternate-duty assignment because of health complications during or after pregnancy. A leave of absence during pregnancy and childbirth (beyond the 12 weeks mandated by FMLA) should be available that will accommodate difficult pregnancies and recovery from a delivery that involves complications. Such leaves usually are unpaid, but they should include continued health insurance coverage.

Many fire department policies require the opinion of a doctor, often the employee's personal physician, to determine how long a pregnant woman can work safely in her fire service job. Since many physicians are not familiar with the actual demands and hazards of firefighting, they should be educated about the job before they are asked to give such an opinion. Fire departments may wish to develop a standard physician's release form for this purpose that specifically lists the requirements of the job. Any such form used in cases of pregnancy must also be used in a comparable way for other nonduty-related disabilities, such as off-the-job injuries.

Parental Leave. Most women will be physically capable of returning to full firefighting duty within 6 to 8 weeks following the birth of their babies. However, a new mother or father may wish to spend more time with an infant beyond the time needed for physical recovery. Prodded by State and Federal law, employers now offer leave to employees who want or need more time to be full-time parents to a new baby or a newly adopted child.

Many employers, including city or county governments and fire departments, have recognized parental leave as a policy that can benefit both workers and their employers. Parental leave policies that surpass the requirements of the FMLA and State laws may be negotiated into contracts, developed by managers as standard practices, or enacted as local ordinances. Four months is a common length for unpaid parental leave in the U.S., although some employers provide up to a year. In an increasing percentage of cases, such leave may also be used, like FMLA leave, to care for an ill or bedridden child or older relative.

Education

The fire department should provide education as part of its reproductive safety policy. All employees must understand the hazards firefighting poses to reproductive health. A qualified physician or other professional who is well versed in the existing research on the issue should conduct classes on this subject for all firefighters and officers. Women firefighters also should be educated early in their careers about the options that exist for them should they become pregnant.

Conclusion

Reproductive hazards and family needs pose manageable challenges to fire departments, just as they do to other employers. Fire departments that continue to view reproductive safety from the narrow perspective of pregnancy (i.e., as a "women's problem") do a disservice to their employees. Employers that do not implement policies adequately addressing employee pregnancy cannot be said to have truly accepted women's presence in the workplace.

Many fire departments have arrived at good solutions that balance the needs of the employee and the employer. A policy that allows the employee to continue as a contributing member of the department during pregnancy, and that assures her of continued pay, benefits, and seniority, is an attainable goal for fire departments of all sizes.

Notes:

[1] Willing, Linda, "Maternity Survey. "*Women in the Fire Service Quarterly*, Summer 1987, pp. 1-6.

[2] Marzella, Louis, M.D., Ph.D., et al. "Carbon Monoxide Poisoning." *Practical Therapeutics*, Vol. 34, no. 5, pp. 186-194.

[3] Olshan, Andrew F, et al. "Birth Defects Among Offspring of Firemen." *American Journal of Epidemiology*, Vol. 131 no. 2, p. 312.

[4] McDiarmid, Melissa, M.D., et al. "Reproductive Hazards of Firefighting I and II." *American Journal of Industrial Medicine*, 19:433-472 (1991).

[5] Milunsky, A., et al. "Maternal heat exposure and neural tube defects." *Journal of the American Medical Association*, 268:882-885, 1992.

[6] McDiarmid, *op cit.*

[7] "Firefighter Reproductive Health Data." *Women in the Fire Service Quarterly*, Summer 1997, p. 2.

[8] Floren, Terese M. "Health Problems Raise Questions for NYC EMT's." *Firework*, April 1993, p. 1.

[9] "Pollutants in Human Breast Milk." *Human Toxic Chemical Exposure*, undated.

[10] McDiarmid, *op cit.*

[11] 42 U.S.C. §2000e (K).

[12] *EEOC Questions and Answers on Pregnancy Discrimination*, C.C.H. ¶3951 (April 20, 1979).

[13] *California Federal Savings & Loan v. Guerra*, 107 S.Ct. 683, 42 F.E.P. 1073 (1987).

[14] Supreme Court of the United States, No. 89-1215. *UAW v. Johnson Controls Inc.*, majority Court opinion delivered by Justice Blackmun on March 20, 1991.

[15] "Those surveyed were asked to describe...the ideal maternity policy. The vast majority outlined a policy that would provide light duty for the term of the pregnancy...then leave permitted during actual delivery and for three to six months after birth." Willing, Linda, "Maternity Survey," *Women in the Fire Service Quarterly*, Summer 1987, p. 4.

[16] Women in the Fire Service, Inc., unpublished survey data, 1995.

Family and Medical Leave Act

The FMLA of 1993 expanded sick leave and family-care leave benefits for millions of U.S. workers. It guarantees employees up to 12 weeks a year of unpaid leave for the birth or adoption of a child or the placement of a foster child, or for a serious health condition of an employee or his/her spouse, parent, or child.

Many States already had family leave laws on the books before 1993. The FMLA is notable not only because it extended this benefit to workers in States that did not have such provisions, but because it requires that health care coverage of employees be continued during such leaves. Most State requirements do not have this provision.

The FMLA provides minimum guarantees and does not take away benefits provided through employer policy or collective bargaining agreements. Thus, if an employer already is required, by contract or State law, to provide more family and medical leave than the FMLA mandates, the FMLA does not reduce that requirement. Employers must comply with whichever provisions are most generous to the employee. Correspondingly, collective bargaining agreements may not be used to diminish workers' rights under the FMLA.

The FMLA applies to all public-sector employers subject to the Fair Labor Standards Act and Federal minimum-wage laws, and to all private employers with 50 or more employees. Employees must have worked for the employer for at least a year, and must work an average of 25 or more hours a week, to be covered.

Key provisions of the FMLA:

- Workers may take up to 12 work-weeks of unpaid leave during any 12-month period in order to care for a newborn child, a newly adopted child or a newly placed foster child; to care for a spouse, child or parent with a serious health condition; or due to a serious health condition that leaves the employee unable to work.

- The employer must continue the employee's health care benefits during FMLA leave. The employer must maintain employee coverage under a group health plan if the employee would have been eligible for such coverage if he or she had not been on leave. (The law does not address the issues of other benefits such as pension contributions during leaves.)

- If employees normally pay a portion of health plan costs, they must continue to pay this portion during leaves. If an employee chooses not to return to his/her job after a leave, the employer can, under certain conditions, demand repayment of the health care premiums it has paid on behalf of the employee.

- When an employee returns from leave, the employer must put her or him back into the position previously held, or into an equivalent position with equivalent pay and benefits. The employee must receive any unconditional pay increases that occurred during the leave, as well as certain other pay bonuses. "Key employees," defined as the highest-paid 10 percent of the workforce, may be denied reinstatement in certain cases if this would cause substantial economic harm to the employer.

- Leave taken for the birth or placement of a child may be taken any time in the 12 months that follow the date of birth or placement.

- "Intermittent" or "reduced-schedule" leaves are permitted for birth or placement leaves; they must be made available for health leaves. This means if a worker has a medical need for small chunks of leave--for example, to take a family member to and from the hospital for weekly medical treatments--or needs to work on a reduced schedule (such as one that allows him or her to be home at night), the employer must make these leaves available. The employer may temporarily transfer employees using these leaves to a different position--for example, from a shift schedule to a 40-hour job--to better accommodate the needed leave, as long as the new position has equivalent pay and benefits. Employers are not required to allow intermittent or reduced-schedule leaves for the birth or placement of a child, but they may agree to do so.

- Employers may require employees to use their paid vacation and sick leave as part or all of the 12 weeks of leave mandated by FMLA. Compensatory ("comp") time given in exchange for Fair Labor Standards Act overtime compensation is not considered a form of accrued paid leave. An employer can therefore not require an employee to use comp time as part of an FMLA leave, although employees may do so at their own discretion.

- Employers do not have to allow workers to accrue seniority while on leave. FMLA leaves do not constitute a break in service for purposes of pension vesting.

- Spouses who work for the same employer may be limited to a combined 12 weeks of leave in the case of childbirth, adoption/placement, or family illness (of someone other than the employee).

- Employees may be required to provide 30 days' notice of their intent to take leave for childbirth or child placement, or for other foreseeable reasons.

- The employer may require a doctor's certificate to verify the need for leave taken for health reasons. The employer may, at its own expense, require a second opinion and, if the two conflict, a final and binding third opinion. Employers may also require periodic recertifications and reports on the employee's status.

- Employees are not eligible for unemployment benefits during FMLA leaves.

- An employer may not restrict the number of employees who may be on FMLA leave at one time.

Definitions:

"Serious health condition" means an illness, injury, impairment, or physical or mental condition involving inpatient care (in a hospital, hospice, or residential health care facility), or continuing treatment by a health care provider. It also includes treatments for various diseases and for surgery to repair injuries. Short-term illnesses are not covered, and common illnesses such as colds and flu are specifically excepted. Employees using leave for their own health conditions must be unable to perform their job functions, though employers also must approve intermittent leaves for employees to receive necessary treatments for early stages of diseases such as cancer.

A "chronic serious health condition" is one that requires periodic medical treatment and continues for an extended period, either continuously or as a series of episodes. Employees who suffer from chronic serious health conditions, or whose family members do, are eligible for FMLA leave, even if they are incapacitated for fewer than 3 days and do not visit a health care provider.

"Son or daughter" means a biological child, adopted or foster child, stepchild, legal ward, or other child under 18 for which the employee stands in the place of a parent. Sons and daughters 18 or older are included if they are incapable of caring for themselves due to a mental or physical disability.

"Parent" is defined correspondingly; the FMLA does not, however, mandate leaves to care for a nonmarital partner or for the parents of one's spouse. (Some State laws do mandate leaves for these family members.)

"Care" given by the employee need not be physical, but may refer to psychological care in the home as well as during inpatient stays in health care facilities. The definition of "health care provider" was expanded in 1995 to include social workers as well as any providers recognized by the employer. The employer's health care provider is permitted to contact the employee's health care provider in order to clarify information in the medical certification, but such contacts may not include a request for additional information about the employee's condition.

"Equivalent position" does not mean that the employee's new position is merely comparable or similar to the one she or he left. The job duties and all terms, conditions and privileges of employment must correspond to those of the original position.

Employers are required to post notices in the workplace explaining the provisions of the FMLA. Most employers will want to go farther, including information about FMLA leave in employee handbooks or other material about benefit programs. Employers who violate the FMLA are liable for the wages and benefits that the employee loses, or for any actual monetary losses. Employers who have not acted in good faith are liable for double damages.

For more information, see the Family and Medical Leave Act, 29 U.S.C.A. §§2601-2654 (1993) or the Final Rules, Federal Register, January 6, 1995 [60 Fed. Reg. 2180-2279 (1995)].

Child care for the fire service

Firefighters have unique requirements when it comes to child care. Most need something other than the normal daytime care options that are commonly available. Firefighters' children often need nonparental child care for more than 24 hours at a time. Many firefighters are subject to emergency call-back during major incidents, and may need someone to take care of their children on a moment's notice at any hour of the day or night. Two-firefighter couples or single parents are especially affected by these circumstances. The problem particularly affects women firefighters: 31 percent of women firefighters are married to or involved with other firefighters; 11 percent are, or have been, single parents.

Many cities and private employers are beginning to take an interest in the child care needs of their employees. They recognize that child care problems cause absenteeism, poor productivity and morale, and may lead to the loss of good employees. Problems with child care are a significant reason why women may not enter the fire service, may leave early in their careers, or may not return after having children.

Although governments and private employers are giving more attention to this issue, most existing programs address only the needs of employees who work relatively conventional hours. The few employers who support child care centers for their workers most often maintain these centers only during extended business hours.[1] Centers that can accommodate 24-hour child care or emergency drop-ins at any hour are virtually nonexistent. Other management practices that may help employees with child care problems, such as job-sharing and flexible hours, either are not feasible or are not made available for firefighters.

Creative solutions are needed, and have begun to emerge. One fire chief has suggested that his city develop old fire stations into around-the-clock child care centers specifically for the benefit of firefighters' children. "We found that some qualified people do not apply for the job because they're concerned about what they'd do with their kids," he said. "I don't think it benefits the department if qualified people get away because of that."[2] In the U.K., the London Fire Brigade provides child care allowances to parents who pay babysitters, and subsidizes spots in child care centers for its employees' children.[3] The Suisun City, California, Fire Department has increased the off-duty response of its full-time personnel to major incidents by outfitting and staffing one of its rehab vehicles to handle child care. Firefighters responding to the call can drop off their children at the fire station or at staging, to be cared for by members of the rehab team for the duration of the incident.[4] Other fire departments provide child care during off-duty training sessions, to encourage greater participation. Positive examples also are being set by some hospitals and airlines, which have many employees who need child care at unusual hours.

As single-parent families and two-firefighter couples become more common, the issue of child care for firefighters will become increasingly important to fire service managers. Problems with child care can prevent individuals from entering the fire service or may influence their career tracks once they are there. The burden falls most heavily on women with children, especially single mothers.

Notes:

[1] Only two percent of all government and private employers sponsor day care centers for their employees' children, according to the U.S. Department of Labor's Bureau of Statistics (1988).

[2] Bowers, Karen. "Old firehouses may get rebirth," *Rocky Mountain News*, August 18, 1989; quoting Denver Fire Chief, Richard Gonzalez.

[3] Allcock, Ann. "Workplace Child Care and the London Fire Brigade," *WFS Quarterly*, Summer 1991; pp. 8-9.

[4] Stevens, Larry H. "But who will watch the kids?" *Fire Chief*, April 1992, pp. 113-114.

Nepotism and firefighter marriages

Fire departments historically have had a strong family tradition. Sons have followed fathers into fire service careers for generations. Brothers have served side by side as career or volunteer firefighters. Now that women are an integral part of the fire service, a new family tradition has emerged. Although women may follow their fathers (or mothers) into the fire service, and sisters and brothers may both choose firefighting vocations, the most common type of familial relationship between male and female firefighters is that of marriage.

According to Women in the Fire Service's 1995 survey, 23 percent of all women firefighters surveyed were married to firefighters. Another 12 percent were involved in permanent or long-term relationships with other firefighters. Most of the marriages were between firefighters on the same department, and the majority of marriages and relationships had developed after both employees were hired.

It should not be surprising that these figures are so high. Once a person has finished school, the workplace is the most likely place to meet friends, social partners, or mates. This is true in any profession, but seems especially so with firefighters. The nature of the job, particularly the unusual hours and work environment, can make a conventional social life difficult. Strong bonds of friendship and loyalty always have developed among firefighters who work closely together under difficult circumstances. It is natural that these same feelings would develop between men and women firefighters, and that good friendships might potentially lead to further involvement or commitment.

Fire departments have reacted to this trend of firefighter relationships in widely varying ways. Some have taken a completely laissez-faire attitude and have not interfered in any way with firefighter couples, even allowing them to continue working in the same station together. Other departments have taken the opposite approach, attempting to impose very restrictive policies on these employees. An extreme case of this type of policy was one that prohibited marriages between employees, and made marriage to another firefighter a cause for dismissal. The policy was made effective retroactively so that the one woman on the department (who was married to a coworker) would be affected. Another example would be of departments that attempt to discover which couples are dating or otherwise emotionally involved with each other, and reassign one or both parties. Such policies go beyond legitimate employer concerns into the realm of harassment, resulting in unnecessary stress for the employees as they attempt to conceal relationships and even marriages from their coworkers.

Policies against nepotism (favoritism to a relative) are a legacy of 19th-century political "spoils" systems and were originally put in place to prevent local political officials from appointing their relatives to jobs. The policies are more common in some parts of the country than others; of fire departments recently surveyed, less than a third had a formal antinepotism policy. Some policies include strong statements such as the following:

> The employment of relatives in the same organization tends to have a number of undesirable results. In the interest of preventing potential abuses in hiring, supervisory authority, or the appearance thereof, it is the policy of the City to limit hiring and supervision of relatives by City employees.

Others take a more moderate approach:

> The department recognizes that there are many situations where two individuals who have a personal relationship may appropriately be allowed to work in the same program, activity or location without adverse impact. However, under circumstances where work, safety, morale, or impartial supervision is demonstrably and adversely impacted by a personal relationship, the affected employees may be accommodated by the reassignment of one or the other.

The antinepotism policies of some cities restrict the hiring of relatives but do not address the potential for employees to become related through marriage once both individuals are employed. Faced with policies that mandate employment restrictions for relatives, firefighters have felt forced to hide relationships or even lie about them. Since some nepotism policies seem to punish people for marrying, couples may choose to cohabit without legal bonds.

Some departments have tried to apply strict nepotism policies to department members who cohabit, assuming the nature of the private relationship that may exist. This is dangerous ground to travel. If cohabiting couples are included in nepotism policies, how about couples who are only dating? What about department members who were once involved but are no longer? How about best friends or off-duty business partners? Will same-sex couples be included in the policy? How does the department know if two people are intimate or merely roommates? Is it really the mission of a fire department, or any employer, to make these kinds of judgments?

The law relating to nepotism is nebulous at best. Rather than holding fast to a restrictive and possibly outdated policy, a department might do better to consider the goals and effects of nepotism policies. Are these policies written to address real problems, or are they punitive? Are policies applied evenhandedly, or are couples ostracized while other relatives are left alone? What specific problems have actually arisen in the department from the employment of relatives or couples?

When restrictive nepotism policies are applied to married couples in the fire service, women firefighters are disproportionately affected. For example, on a department of 300 members with 10 women firefighters, 3 of those women are statistically likely to be involved with other firefighters. Although their partners would represent only one percent of male firefighters, the women involved represent 30 percent of all women on the job. If a department enacts a restrictive policy limiting employees' career opportunities because of their relationships, that department is accepting that one in three of its women firefighters will be limited in this way.

In times of downsizing, few organizations can afford to throw away an individual's contribution. Yet this can be the effect when restrictive nepotism policies are applied when no actual problems exist. The original purpose of nepotism policies was to ensure that employees or applicants would be judged by their own merit and not by who they were related to. Just as employees should not be advantaged unfairly because of their personal relationships, neither should they be unfairly disadvantaged.

Relationships in the workplace are a fact of life. They need not be viewed as a problem to be solved or as a situation to be prevented. Where real problems do exist, either with the couple themselves or due to co-workers' or managers' resentment of the couple, they should be dealt with individually. Perhaps the best policies are ones that take into account job performance as they seek to manage relationships in the work environment. It is possible to develop policies that both respect the rights and privacy of individuals and at the same time maintain a professional work environment for all employees.

Fire station facilities

Most fire stations in use today were planned and built with a single-gender workforce in mind. Many of these buildings now are being used by a workforce that includes both women and men. Not surprisingly, this can result in inadequacies that are a source of inconvenience, discomfort, embarrassment, and friction for all concerned.

Different fire departments have developed a variety of solutions to the problems created by inadequate facilities. The cheapest and easiest answers are usually the first to be implemented: a "men/women" or "occupied/unoccupied" flip sign on the door of the station's only restroom or shower, or a lock on the door, can be installed readily. Makeshift partitions, such as a row of lockers or a rollaway curtain, often can be put up between beds if bunkroom separation is desired.

These are short-term solutions to real or perceived needs concerning personal privacy in the fire station. The underlying question that guides the development of long-range answers is whether men and women on the job should be provided with separate facilities or not.

The legal background

The expense to the employer of providing separate restrooms, showers, locker rooms, and bunkrooms for women and men usually will not support excluding women from an occupation.[1] According to the EEOC's guidelines,

> Some states require that separate restrooms be provided for employees of each sex. An employer will be deemed to have engaged in unlawful employment practice if it refuses to hire or otherwise adversely affects the employment opportunities of applicants or employees in order to avoid the provision of such restrooms for persons of that sex.[2]

An even more emphatic provision occurs in the Guidelines of the Office of Federal Contract Compliance:

> The employer's policies and practices must assure appropriate physical facilities to both sexes. The employer may not refuse to hire men or women, or deny men or women a particular job because there are no restrooms or associated facilities, unless the employer is able to show that the construction of the facilities would be unreasonable for such reasons as excessive expense or lack of space.[3]

In one case where a firm had refused to hire a woman welder on the grounds that its repair yard lacked locker and restroom facilities for women, the EEOC discovered that the employer had actually had separate facilities during World War II when it had many women workers.[4] Since the existence of male-only facilities is often the result of the past discrimination that Title VII was designed to eliminate, allowing cost as a defense would only honor and perpetuate that discrimination.

One State law that applies to some fire stations is Section 2350 of the California Labor Code. It requires that business establishments that have five or more employees must provide separate bathrooms for each sex, and that no person may use bathrooms designated for the opposite sex. Other States may have comparable provisions in their labor codes or other laws.

Local building or health codes usually require employers to provide bathrooms, and sometimes other facilities, for each gender in the workplace. As most fire stations are the property of municipal or county government, they generally have been made exempt from the provisions of these codes. Where such exemptions do not exist, of course, the fire department would be responsible for compliance.

A few fire departments still assign women only to stations that "have facilities for women." This is not an acceptable long-term solution, particularly where stations assignments are on a seniority bid basis and the woman otherwise would be entitled to bid for the off-limits stations. The result can be an unworkable inflexibility for management. It can also generate resentment from male coworkers that the woman doesn't have to take her turn at roving, particularly if her low seniority would normally make this part of her job. And it is unfairly restrictive of the woman, if she is not permitted to make shift exchanges or time trades with firefighters at certain stations.

The impact of inadequate facilities

One fire service observer, commenting on the problem of inadequate station facilities, wrote

> Under the best circumstances, bad facilities are an inconvenience which women suffer from in far greater proportion. Under the worst conditions, poor facilities can lead to problems with morale and job performance, and an increase in the occurrence of harassment. At least one discrimination lawsuit has been filed which was due in part to inadequate facilities. A lawsuit costs a lot more than a locker room, and in the end, no one wins.
>
> When the need for women's facilities in the fire station is neither recognized nor addressed, the... department may be saying that women are not important enough here to deserve decent facilities, that women may not be around long enough to warrant planning for the future, that women are not wanted at this station, and this is a reasonable way to keep them out; or that we are too busy here to consider the real needs of our personnel. All of these are harmful messages, both for women and for the organization of which they are a part.[5]

Two things commonly happen when firefighters in a newly integrated workforce are forced to occupy inadequate facilities. One is that the women are blamed for "causing" the problem. Even though it is the design of the station that is lacking, the feeling among the men is often that since "there wasn't a problem until she got here," it's the woman firefighter's fault. Solutions such as bumping an officer out of his private room to give it to the woman can generate a similar resentment. Where this type of hostility exists, providing facilities that offer privacy for both genders becomes only half of the solution. It is important for management to make it clear that alterations to the facilities are being done not "for the women" but for better privacy for women and men alike.

Another common reaction to inadequate station facilities is the tendency to adapt and accommodate to them. Women and men in the workforce--and particularly women, if they are in the minority or are the newest firefighters--usually will adapt to situations that are less than ideal. Many women firefighters do not routinely shower at work, or they get up an extra hour early in the morning in order to shower before the men need the shower facilities. Women have learned to use broom closets as changing rooms; firefighters of both sexes develop the habit of looking for feet under the restroom stall walls. But just because a person or group can develop behaviors to cope with a situation or environment doesn't mean it's right to leave things that way indefinitely. All firefighters deserve basic privacy, either individually or by gender, for personal functions. Management should make it a priority to provide this.

Solutions

Develop a 5- or 10-year plan for remodeling your firehouses. All new stations and any significant remodeling of existing stations should include adequate facilities for a two-gender workforce. Most fire stations will stand for a half century or more. To continue to build them on old designs means you will perpetuate the problem fire station designers 50 years ago created for you today.

Although crews in many fire stations manage to cope with shared facilities, it is preferable for station design to provide privacy for both sexes in restroom, shower, and changing areas. The issue of separate bunkrooms for women and men is more controversial. As mentioned earlier, reassigning an officer's room to a woman firefighter usually creates hard feelings. Tucking an ad hoc "women's bunkroom" off in one corner of the station (such as a rollaway bed in the weight room) is inconvenient for everyone and a clear message that the woman doesn't belong. The most common solution is for women and men to share the one existing bunkroom. Many women firefighters prefer this arrangement, because it keeps them a part of the crew and a part of the information-sharing process that begins as soon as a call comes in. On the other hand, some men and women are not at all comfortable sharing a bunkroom in this fashion.

The real long-term solution to the bunkroom question is to provide privacy for everyone. Many new firehouses are now being built to a design that features cubicles containing a bed, desk, lamp, and three or four lockers (for one person on each shift), with a curtain across the doorway. This provides privacy and a reduction of sound or light from the others in the room. (Snoring may be a common source of humor among firefighters, but routinely being deprived of sleep by one or more snoring coworkers is also a significant source of job-related stress.) If the partitions do not extend all the way to the ceiling, the open space at the top allows for air circulation and allows emergency tones and information to be heard.

This is a solution that pleases everyone and doesn't pit the women against the men, the paramedics against the suppression personnel, or the officers against the firefighters. It also avoids controversies over whether the women's bunkroom in the new station should be the same size as the men's bunkroom, whether men are allowed to use the women's bunkroom if no women are assigned to the station on that shift, whether a station that houses two officers should have men and women officer's bunkrooms, and so forth. It is a solution that respects the privacy and individuality of all firefighters without regard for gender, and for that reason is usually supported by all concerned.

Notes:

[1] As a rule, expense will not support gender as a bona fide occupational qualification unless the expense would be clearly unreasonable. See EEOC Case No. YNY 9-047 (5/21/69), C.C.H. Empl. Prac. Guide ¶6010.

[2] 29 C.F.R. §1604.2 (b)(5).

[3] 41 C.F.R. §60-20.3(e)(1970).

[4] EEOC Dec. No. 70-558 (2/19/70), C.C.H. Empl. Prac. Guide ¶6137.

[5] Willing, Linda, "Bedrooms and Bathrooms: The Hidden Message," *WFS Quarterly*, Winter 1988-89, pp. 1-2.

Hair and grooming standards for firefighters

As the fire service begins to break away from a traditional, authoritarian approach to its employees, fire service managers have begun adopting more flexible approaches to employee grooming standards. A fire chief who takes an extreme approach, such as requiring all employees to have military-style short hair based on male norms, is demonstrating the department's hostility to the presence of women and will find recruitment and other diversification efforts difficult.

This section of the manual explores the legal background for firefighter hair-length requirements, from the conflicting perspectives of the EEOC and the courts. It will discuss the two primary reasons for fire departments to have hair-length policies, and offer guidance in policy development. Although this section does not explicitly address related grooming issues such as the wearing of jewelry, some references to jewelry are provided in the sample policy language on page 85.

The legal background

The courts generally have allowed employers more latitude in establishing work rules for hair length than they have in dealing with unchangeable characteristics such as height. Many early hair-length cases were brought by men, including police officers and firefighters, challenging standards which prohibited long hair on men. Since the mid-1970's, the courts have consistently held that prohibiting long hair for male employees is not sex discrimination under Title VII if the employer's grooming standards for both sexes are related to community standards and are applied in an even-handed manner.[1] The courts have made a distinction between hair standards and issues where discrimination affects "fundamental rights" or is based on immutable (unchangeable) characteristics.[2]

The judicial and administrative branches of the government disagree on a basic point of sex discrimination law as it affects hair and appearance standards on the job. The courts have defined Title VII as a weapon against sex-linked practices that seriously impair fair employment, but have decided that the law was not intended to interfere with employers who wish to set reasonable appearance standards required by their business. The Equal Employment Opportunity Commission (EEOC), on the other hand, has consistently treated different appearance rules for men and women as constituting sex discrimination under Title VII.[3] At the same time, the EEOC has upheld reasonable concerns over safety hazards in industry.

While the applicability of Title VII to hair standards is thus undecided, it is clear that the constitutional protections afforded to employees in this area are limited. Numerous cases have been brought on constitutional grounds, with employees arguing that appearance is an aspect of personal liberty, so that any interference with that "right" must be justified by a legitimate State interest. The courts, in response, have generally upheld a public employer's right to impose grooming regulations that can be justified by the need for discipline, uniformity, and *esprit de corps*, even where no safety considerations exist.[4]

A 1992 case filed under Louisiana constitutional law, however, resulted in a ruling against a public employer. At issue was the city's recent attempt to restrict firefighters' hair to shoulder-length or shorter. (Women had been on the fire department for 11 years and previously had been permitted to pin their hair up to conform with the hair standard.) In issuing an injunction against the city, the judge found that "the regulation (was) not gender neutral" because it affected women differently from men and "ha(d) the effect of classification of individuals on the basis of sex." The ruling continued

Similarity of appearance can be and was in fact achieved by requiring fire personnel on duty or in uniform to have their hair, if longer than regulation length, pinned to meet a length requirement established by their employer. The esprit de corps and equal treatment as to grooming standards are easily achieved by uniform enforcement of hair length **while on duty** or in uniform with a recognition of an individual's right to have whatever length of hair he or she desires **as long as while on duty or in uniform it is kept to a level set forth in the employer's regulation**...While similarity of appearance has been recognized as an appropriate and rational goal in a "para-military" civilian service, commonality may not and should not be required at the expense of reason and purpose.[5]

Safety versus grooming

Apart from the legal aspects, confusion also exists on hair-length issues because fire departments have, in the past, required firefighters to have short hair for two different reasons: because, given the protective gear then in use, it was safer during firefighting operations; and because it looked neater, more professional, or more uniform. Where these two reasons have become intertwined into one policy, it is necessary to sort out the real and justifiable reasons for any given policy.

A safety standard is gender-neutral: fire will burn exposed hair regardless of the person's sex. Advances in protective equipment in recent years have made it possible for firefighters to have much longer hair than in the past and still be much safer than ever. SCBA facepieces and flame-resistant hoods and helmet liners ensure that all surfaces on the head are protected. For this reason, many fire departments have adopted a safety-based hair standard that simply states that "No hair shall be exposed during firefighting activities." This approach avoids the need to regulate the actual length of the hair, and bases its restriction simply on the justifiable need for firefighter safety.

Establishing a policy that is based solely on grooming and appearance concerns has a more favorable impact on women, for while a safety standard must be gender-neutral, a grooming standard need not be. Even though the EEOC disagrees, the courts have consistently held that employers can legally require male employees to wear their hair shorter than female employees and that such policies do not constitute illegal sex discrimination.

Policy development

Firefighter hair length easily can become a controversial issue within a department. From a commonsense perspective, even if not a legal one, finding a solution requires a balancing of employee and employer interests. For the employee, hairstyle is a matter of personal identity, and a preferred hair length cannot be put back on at the end of a work shift in the way that a uniform is removed. The employer's responsibility for personnel safety and their interest in a professional appearance can encompass a wide diversity of hair lengths and styles. The best grooming standard may well be one that screens both sexes on a community-based standard of dress and appearance, a standard that applies an equal burden to both sexes.

Notes:

[1] *Dodge v. Giant Food, Inc.*, 488 F.2d 1333 (D.C.Cir. 1973); *Baker v. Calif. Land Title Co.*, 507 F.2d 895 (9th Cir. 1974) cert denied, 422 U.S. 1046 (1975); *Longo v. Carlisle DeCoppet & Co.*, 537 F.2d 685 (2d Cir. 1976); *Earwood v. Continental Southeastern Lines, Inc.*, 539 F.2d 1349 (4th Cir. 1976); *Barker v. Taft Broadcasting Co.*, 549 F.2d 400 (6th Cir. 1977); *Knott v. Missouri Pac. R.R. Co.*, 527 F.2d 1249 (8th Cir. 1975).

[2] E.g., an employer's refusal to hire women with small children [*Phillips v. Martin-Marietta Corp.*, 400 U.S. 542 (1971)] or the firing of women who marry [*Sprogis v. United Airlines*, 444 F.2d 1194 (7th Cir. 1971)].

[3] E.g., EEOC Dec. No. 71-1529, 3 F.E.P. 952 (5/9/71).

[4] *Quinn v. Muscare*, 425 U.S. 560, *reh'g denied*, 426 U.S. 954 (1976) [male firefighter challenge to hair standard]; *Kelley v. Johnson*, 425 U.S. 238 (1976) [male police officer].

[5] *Sellers v. City of Shreveport, et al.*, 1992. The employer defended the policy as a grooming standard, stipulating that the safety of firefighters was not an issue.

Sample language from fire department hair-length and grooming policies

Following are excerpts from fire department policies regulating employee hair length and the wearing of jewelry. They cover a range of options available to fire service employers for dealing with these questions. No such policy should be implemented without considering your own department's specific needs or seeking qualified legal advice.

"Hair shall be neatly groomed and the length or bulk of the hair shall not be excessive or present a ragged, unkempt or extreme appearance. (Men:) Hair may cover one half of the ear but shall not cover the entire ear. (Women:) Hair may not extend beyond the lower part of the shoulder blades."

"Members are discouraged from wearing rings or other jewelry on the fire or training ground. Female members may wear earrings providing they do not extend below the bottom of the ear."

(Women:) "Hair must be clean and neatly arranged. When in uniform, back hair must not fall more than one-quarter inch below the lower edge of the collar. No hair must show under the front brim of fire service headgear. Afro, natural, bouffant, and other similar hair styles are permitted, but...bulk of hair must not exceed two inches. In no case is the bulk of the hair permitted interference with the proper wearing of fire service headgear."

"Only pins, combs or barrettes that are similar in color to the individual's hair color may be worn to meet the requirements of the regulation. Jewelry which extends beyond the ear lobe or...is loose or protrudes and may catch in machinery or equipment may not be worn while on duty."

"It is recognized that traditionally acceptable standards for female hairstyles differ considerably from those of males. Female hairstyles that would normally not conform to the standards outlined in this S.O.P. may be pinned up or secured in order to comply while on duty. In these instances, the hair must be pinned up or secured at all times while on duty, and shall not interfere with the proper wearing of uniform hats or protective equipment, or in any way create a safety hazard."

(Men and women:) "There are many hair styles that are acceptable. So long as the person's hair is kept in a neat, clean manner, the acceptability of the style will be judged by these criteria: Hair styles that preclude the proper wearing (of SCBA) are not permitted... Hair will be worn so that it does not extend below the bottom of the uniform shirt collar while standing in an erect position. Hair may be pinned or worn in a way to keep hair above the bottom of the collar..."

"To facilitate a professional appearance, hair and grooming standards must be followed. These standards have been modified to meet contemporary styles without jeopardizing the safety of firefighters involved in the hazardous activities associated with firefighting."

"When in a normal standing position, the hair can extend to the top of the collar area. Hair will not extend beyond the bottom of the earlobe. Longer hair is acceptable if it is pinned up in a neat manner and does not interfere with the wearing of departmental headgear. No ribbons or ornaments shall be worn in the hair except for neat inconspicuous bobby pins or conservative barrettes, which blend with the hair color. Hair...will not exceed two inches in height."

Promotional issues for fire service women

Women firefighters who consider advancing into supervisory ranks face different circumstances than their male peers going for the same promotions. In some cases, women must overcome obstacles that may make them hesitant to try for promotion or may hinder their success once they've been promoted. In other cases, women find their different perspectives and circumstances are an advantage in fulfilling the officer's role. Yet in some key ways, women and men are much alike as they progress through the ranks on their fire departments.

Women in general do face some obstacles that many of their male counterparts do not experience as they advance in their careers. The most obvious factor is that a woman does not "look the part" of a traditional fire officer. Few citizens expect a woman to show up as part of an emergency crew, much less as the leader of that crew. Most people have deep-rooted societal conditioning about who they expect to be in charge. When, big, strong-looking men are on the scene, the average civilian will not assume a woman to be in the position of authority over them.

The result of this is that some women officers find themselves being discounted or ignored by the public. Minority men also have had this experience. This is a demoralizing situation at best, and may be worsened if crew members try to step in as informal leaders. (Their attempt to "help out" may be read as a lack of confidence in her authority). A woman with a supportive, professionally minded crew will have the best success in educating the public about the new roles women play in society.

Women may suffer damaged confidence as leaders if they feel others around them perceive them to be less competent than other, male, firefighters. Again, the public is often to blame for this as they make possibly well-meaning but condescending statements to women that they would never dream of making to men. Most women firefighters have heard, "Oh, you can't be a firefighter; you're too small," or "You don't really go into burning buildings, do you, dear?" In these cases, it is not the size or ability that is being commented on, but gender.

Unfortunately, some male coworkers may share these invalid stereotypes about the capabilities of women, and act accordingly. Poorly trained men may treat a woman officer differently and try to undermine her authority. Such behavior damages the effectiveness of the entire crew. This creates problems for individuals, and also diminishes the quality of service provided to the public.

Training is critical in mitigating or preventing these types of problems. All officers should receive leadership development training and guidance in teambuilding with their crews. All department members will benefit from training in communications and issues of harassment. The woman officer also needs to know she has the genuine support of her supervisors if such problems come up.

Women usually lack the access to the informal side of the organization available to most men. Women are less likely to play on predominantly male sports teams, go hunting or fishing with "the guys," or otherwise socialize informally with their male coworkers. In such settings, where rank is put aside in favor of camaraderie, those present can gain tremendous insight into the subtext of an organization. They also gain important information about coworkers this way. Not having access to these informal avenues of learning can be detrimental. Until there are enough women at all ranks on the job to create networks that are open to women, women may have to find alternate and possibly less effective ways to understand the nature of the organization completely.

Because a sense of membership and acceptance among a small group is so beneficial, women should be encouraged to find alternate sources of this kind of networking and support. Membership on special teams or committees can give women much confidence and a sense of place, particularly on larger departments.

Recruiting women for diverse assignments they may not be aware of, or may be hesitant to try, can help individuals find new career paths and enhance work groups with new outlooks.

One of the greatest obstacles to the first women pursuing promotions is that they have no role models. They may never have seen a woman driving a fire truck or a woman commanding a fire scene. When there is no one who looks like them performing a particular job, it is very hard for most people to imagine themselves in that role. This is one reason why promotions for women firefighters often have a snowballing effect: it may take a long time for the first woman to step forward into a promotion, but after that, increasing numbers of women are likely to follow as they see a real person like them in the position.

Going for promotion feels risky for many women firefighters. They may wonder why they should endanger their hard-won acceptance and confidence as firefighters only to step into unknown territory. In this way, women are not that different from men who also fear being seen as incompetent in new roles and may question if it is worth it. The difference is that women may feel they have much more to lose, and may have paid a much higher price for the security they feel they are leaving behind.

Women may also have some distinct assets as fire service leaders. Women with diverse backgrounds bring valuable skills to the position of fire officer. Women who felt insecure as firefighters because they didn't know how to tear apart a diesel engine suddenly may find that written and verbal communication skills, the gift for organizing, the ability to teach effectively, and other skills of this type are more important in the officer's position.

When a woman realizes that she has abilities that are needed in her new role, she is likely to gain confidence in using those skills and enhancing other traits that may need further development. Knowing she has a solid base in some skills can make her feel more capable of dealing productively with any weakness in others.

Women certainly do not have a corner on intuition, but it is a quality that historically they have been encouraged to develop, and it is an extremely valuable quality for fire officers. High-level technical and skills training combined with sensible regard for intuition can be a powerful combination, whether the challenge is commanding a fire scene or handling personnel problems in the station.

Like each man, each individual woman brings a unique set of insights and experiences to the job. When she is allowed to value her own contribution and encouraged to try new things without fear of devastating failure, she is likely to grow and contribute increasingly through every promotional step she attains. Men and women alike must develop leadership styles over time, and no two fire officers ever will be identical. Women bring different outlooks to leadership roles and will enhance the organization that values them and encourages their efforts to attain roles of increasing responsibility.

Many women hesitate to make the first step toward these positions of greater power and responsibility. Women may find it easier to stand back while their male peers go after leadership roles, telling themselves that their real contribution lies in "support." In fact, many women are happy and successful as firefighters and find fulfillment in retaining that position for an entire career. Women should not be forced to go for promotions; they should not be made to feel guilty if they choose to wait.

On the other hand, a fire service manager may have a number of women on the job with plenty of experience to qualify them for promotion, yet none of them chooses to test for advanced rank. A manager in this situation should try to find out whether the women feel they would be welcomed and supported as officers. Have they faced, or do they continue to face, harassment at work that keeps them down? Do the women feel they lack the training, both within and outside the department? Do the women lack confidence because they have been protected at less active stations or worn out proving themselves at extremely tough (or frequently changing) assignments? Is the promotional rank perceived by women to be unappealing for some reason?

Fire chiefs have a vested interest in making sure their promotional processes are as inclusive as possible. Only the most inclusive, competitive tests can produce the best candidates. Women must be encouraged to participate in these processes, and must feel assured that promotion will bring opportunity and the potential for increased professional fulfillment, not a renewal of old problems with acceptance.

Fire service managers have tremendous power to develop their employee resources in preparation for promotion. In some cases, women may benefit from an individualized focus. In all cases, efforts that encourage all department members to be their best and contribute at the highest level will make for a more professional, smoother running, better performing fire department.

A key component to supporting the promotional success of women is to renew management's commitment to providing a supportive, harassment-free working environment for everyone. Women who feel they face harassment and other inappropriate treatment based on their willingness to try for promotion soon will give up the effort. It just won't seem worth it. This is especially true for women who have faced workplace harassment in the past, and who may still be dealing with it.

Effective leadership and training are essential to eliminate harassment from the workplace. Outside instructors with good knowledge of the fire service can be very helpful in these training efforts. They may be able to see problems more clearly and prescribe objective solutions. All department members should be trained, with emphasis placed on those in supervisory roles. Harassment is a problem that is fed from the top down, and it is critical that leaders understand and model appropriate behavior.

Leadership development training also can be very helpful for aspiring officers. Women may benefit most from this type of training if they are allowed to train in groups with other women. When women discuss leadership as minority members in groups of men, they are likely to allow men to set the standards of the discussion, and it may be difficult for them to raise their own concerns that seem to differ from the interests of the majority. Fearing to look stupid or too much "like a girl," being unwilling to have the discussion focus overly on the concerns of a small minority, or simply having trouble getting a word in edgewise, they may be silenced effectively, and their questions never may be addressed. When students feel they have a safe place to practice skills, ask questions, and make mistakes, they are much more likely to take risks and grow professionally.

Some departments develop a leadership curriculum in house. If this curriculum includes training particularly for women, it is very useful to include a woman fire officer as a primary or adjunct instructor. A woman who has "been there" can answer some of the questions that are not addressed in strict management theory.

Another way to provide leadership development training for women is to support their attendance at training seminars and conferences outside the department. Classes that include people from different fire departments and different parts of the country can create an environment that allows for new approaches and insights. Many women feel energized as a result of these experiences and return to their departments with new commitment and desire to promote.

Conferences and classes especially for women firefighters can provide a doubly enriching experience, as women get fresh insight as well as support from other women and access to role models. Women in the Fire Service, Inc., a national nonprofit organization based in Madison, Wisconsin, sponsors training opportunities and conferences at the national level specifically geared to the woman firefighter and fire officer. Other agencies and jurisdictions sponsor similar events at the local and regional levels.

All women come to the fire service with an unique set of skills that will help them succeed as fire officers, but these skills must be recognized and supported. Fire service managers may find it useful to do skills inventories of all employees, and then work to find ways to develop, use, and share those skills within the department. Frequently women and men alike are shy about offering to do work in new areas, but usually will rise to the challenge if specifically asked to contribute in this way.

Aspiring women fire officers benefit greatly from being mentored, but many have less access to this type of relationship than their male peers. Men with access to power and information about the department might hesitate to groom a woman for promotion as they might groom promising young men. In many cases, this is not even conscious--experienced officers do not see the late nights talking strategy at the coffee table, or the casual golf game, as time spent mentoring. They may feel they are not appropriate to serve as guides for young women, because they fear any interest paid will be misinterpreted as a personal relationship, or that as men they cannot have a full understanding of a woman firefighter's circumstances.

When men hesitate to become mentors to women for these reasons, it usually is because they do not clearly understand the mentor's function. They confuse being a mentor with being a friend or a role model. In fact, men often do a wonderful job mentoring women in the fire service once they feel more clear about the nature of the mentoring relationship.

Because many people do not have a clear idea about what mentors are supposed to do, some departments have instituted programs that encourage mentoring through training and even matching of mentor and mentee. Such programs allow equal access to these types of relationships that are so helpful to a new or aspiring fire officer.

Women are promoting to fire officers' ranks in increasing numbers each year. Women who promote usually find they have increased opportunity to make a contribution, and thus feel increased job satisfaction. This leads to greater commitment to the organization, manifesting in high job performance, greater respect among coworkers and increased longevity on the job. Women who feel supported in their decision to promote, and who have equal opportunities to succeed in their new ranks often will see themselves moving from a fire service job to a fire service career. This is a positive step for both the individual and the department.

Supporting workforce diversity

A hundred years ago, most fire departments in the U.S. included just one group of people: white Protestant men of Northern European backgrounds. For decades, this profile changed very little. Eventually, and not without controversy, Italian-American men (and, later, men from Hispanic backgrounds) gained entry into firefighting jobs. As early as the 1920's in some cities--and very much later in others--a few African-American men became career firefighters. Not until many decades later did their right to do become protected by law, and only in the 1960's did black men become career firefighters in meaningful numbers. In the 70's and 80's, it was women's turn. By the 1990's, several thousand women were working as career firefighters in the U.S. and elsewhere.

This historical perspective makes two things clear. First, the trend is towards ever-greater inclusiveness in the fire service workforce. Second, the speed of change in the composition of the workforce is accelerating. Given these facts, it is not surprising that some members of the dominant group (white male firefighters) should feel uneasy or sense a threat to their longstanding dominance of the workplace and its culture.

Women firefighters are the most recent and perhaps most dramatically different newcomer. Women firefighters represent change. Whether or not they as individuals mean to change anything at all, women are viewed by other firefighters as agents of change to the culture of the fire service. They also alter or threaten the ways male firefighters perceive the job, themselves, and women in general.

Newcomers and assimilation

How have fire departments dealt with the prospect of any new, previously excluded group seeking to become firefighters? The pattern is the same regardless of the group. The first response is usually to build a wall: to find reasons why X's (people belonging to the given group) "can't" be firefighters. Much energy is spent documenting and defending this position.

When enough X's do eventually become firefighters that this argument is weakened, the response shifts: "Okay; we'll hire X's, but we're not going to change anything for them. They'll just have to deal with the workplace as it is." The changes contemplated by this approach include things like restraining firefighters from racist behavior, taking steps against sexual harassment, and providing restroom privacy for female firefighters (and maintaining it for male firefighters). This philosophy is called assimilation. Many fire departments in the 1990's took this approach to women's presence: women are allowed to be firefighters if they will blend into the scenery and not expect any "special treatment," a code phrase for change.

What happens in a workplace when assimilation--the melting pot--is the goal? People who are not part of the dominant group are expected to become as much like the dominant group as possible. A few X's will be able to do this; a few more will be able to do it in limited ways or for a short period of time. For the majority who can't conform in significant ways over the long haul, the results are stress, exclusion, isolation, role confusion, unhappiness, poor performance, and poor evaluations. X's usually are left out of informal communications networks, and find paths to both informal and promoted leadership roles closed to them. (Informal group leaders in many fire crews wield more real power and influence than the officers.)

When assimilation is the goal, it is expected that everyone will have identical needs and be treated identically. Training on affirmative-action and equal-employment issues, instead of being seen as helping everyone adjust to workforce changes, is written off as "special treatment" for the X's. "They didn't have any classes like that for me when I came on," is a common protest.

Networking and support groups also run counter to this philosophy. The assimilation process encourages everyone to try to be like the dominant group; thus, X's are actively discouraged or intimidated from seeking support from other X's. When employees from outside the dominant group connect with others for mutual support, they find they have many issues in common. But identifying issues that exist for X's as a group violates the basic premises of assimilation: X's are not supposed to have issues separate from those of the dominant group. Distrust is often a factor as well: white men may see two or more X's getting together as a potential threat.

Women firefighters face two conflicting sets of constraints. In many ways, they are expected to fit in and become "one of the guys": to be interested in the same things (hunting, fishing, cars), to share the same kinds of humor, to enjoy the same foods, to use the same station facilities comfortably, and to perform firefighting tasks using the same physical techniques as men. But in other ways, women are expected to confirm to the dominant group's ideas of what women (ladies) should be: compliant, subservient, smiling, never angry; and above all, meeting men's standards of physical attractiveness. For women firefighters, these expectations overlap and conflict. A firefighter should be aggressive, but women shouldn't. Fire officers are praised for having a strong command presence and giving orders in a direct and forceful way; ladies aren't. A lady isn't supposed to be around pornography, but women firefighters who seek to have pornography removed from the fire station are often ostracized.

Glossary

Culture: a set of beliefs and values shared by a group of people.

Diversity: variety, differences.

Dominant group: the group of people in a workplace or other setting, usually in the numerical majority, whose values and culture set the standards for everyone there.

Monocultural: dominated by the values of one culture; ethnocentric.

Multicultural: respecting and valuing the cultures of all.

White men set the standards for everyone in the fire service. That this might not be universally beneficial, or that other valuable standards might exist, was rarely considered until recently. But by the mid-1990's, some fire departments had taken a significant step away from the philosophy of assimilation and toward one of valuing diversity. This viewpoint recognizes that the monocultural values of the fire service are increasingly out of place in a diverse work setting. The fire service has changed its workforce without allowing the workplace to change in response. Assimilation uses the melting-pot analogy as its ideal: all people are thrown in, melted down and made the same. Valuing diversity, on the other hand, uses the mosaic: a picture whose beauty comes from its thousands of tiny tiles of different shapes, sizes, and colors.

Assimilation fails as a strategy for many reasons. It creates barriers for the newcomers, wastes talent and creativity that could have benefited everyone, and artificially reinforces the values of the dominant group by making it appear that they are held by everyone. When new employees who differ from the dominant group are not allowed to express their differences, only the most adaptable or invisible will stay on the job. The result is a high turnover of employees, and a loss of much of the value of having hired a diverse workforce. Members of the dominant group, for their part, believing their ways to be the only ways or the best ways, will feel threatened by even the small inroads made by members of other groups.

Valuing diversity works because it pays attention to what the work environment is like for everyone in it. It makes visible the informal workplace support systems the dominant group takes for granted, and recognizes that people who are excluded from these systems are, in fact, getting a negative form of "special treatment." Hiring a few women or other X's and dropping them to sink or swim in the white male culture of the fire station is not a way to manage change progressively. Positive leadership in workforce diversity means working to build a fire service culture where all employees can function productively together.

Fire service leaders speak out on diversity

Why do you value diversity in your organization and in the fire service, and why is that diversity not only necessary but a benefit in carrying out your organization's mission?

Al Whitehead, President, International Association of Fire Fighters:

"The IAFF believes that it is through the collective strength of its diverse membership that it has fairly earned the honor and respect necessary to move forward and accomplish its mission...The success of the IAFF hinges on the diversity, strength and dedication of its members."

Frank S. Rivera, Division Chief, Metro Dade County, Florida, Fire Rescue:

"Diversity, in terms of ethnic and/or gender, racial, or religious makeup is important in order to maintain a department which mirrors our community. If these balances can be maintained, the perceptive problems which usually crop up within departments which don't strive (for) those goals can be avoided. The end result is high morale and productivity."

Jeanne Pincha-Tulley, Forest Fire and Aviation Management Officer, Mendocino National Forest, California

"Many things from the past are good practices, the basis for a firm foundation. Diversity will provide the bridge to the future. It's not an easy proposition to get all the players involved to where they value and nurture diversity; not yet. This is an age of transitions, and true to form, most change is two steps forward and one backwards. We all hang onto concepts and traditions that hamper smooth progress towards diverse thinking and blending ideas, towards treating everyone like a valued human being."

James R. Cavellini, Assistant Chief, Management Services, San Francisco Fire Department:

"It is important that our Department reflect the make up of our community...providing all members of our community a place which provides security and personal regard. As an organization we benefit from the greater respect and spirit of cooperation our community affords us as they recognize our membership is open to them and their families and friends."

Chief Manuel Navarro, Colorado Springs Fire Department:

"I value diversity from both a personal perspective and a community perspective. I personally believe that this country was founded on a basic principle; that is, in the United States we believe and promote your right to be different. To exclude anyone from full participation would run counter to community norms. Diversity is (also) of great assistance in carrying out the mission of the organization. Fire service organizations are simply that: service organizations. Whether the service requires special attention because there are ethnic, religious, racial, sexual preference, or gender-related issues, a diverse workforce enables fire chiefs to manage and provide the highest quality of service."

How have you addressed diversity issues within your organization, and what changes have come from this?

Al Whitehead:

"Discrimination and harassment can best be stopped in both its open and hidden forms by encouraging all people to live by the basic principle of respect and that every person is entitled to the same basic rights regardless of various differences. The IAFF firmly believes that it cannot operate effectively as an organization unless it commits itself to eliminating the practice of discrimination and harassment in the fire service. IAFF members elect twelve members to form an active Human Relations Committee which assists in the handling of various human relations issues. Furthermore, the IAFF conducts an annual Human Relations Conference which directly addresses current human relations issues within the fire service and the IAFF."

Jeanne Pincha-Tulley

"Changing a workforce entails more than the addition of a different gender and minorities. It encompasses a change in thinking. Just bringing in a few 'new people' does little more than set the stage for tokenism and strife. Real change begins with creating the climate, dealing with new ideas and positively reinforcing new thinking."

James R. Cavelini

"Recruitment plans have been developed to provide a greater number of applicants from under-represented groups. Ongoing sensitivity training is given in an effort to overcome concerns of all members...We encourage and recognize employee organizations so the needs and concerns of all members are addressed. The most important change is the cooperation and openness seen in the interactions of the members of our Department."

Manuel Navarro:

"I serve as a role model to other members of the organization; I articulate a vision of a diverse organization, and I manage the organization to model that behavior. In modeling behavior, it is important that I and other leaders of the organization "talk the talk and walk the walk." You cannot lead an organization to embrace diversity unless you demonstrate that you value diversity. In all organizational issues, I engage employees in an open and non-threatening dialogue on diversity. The benefits of diversity, how diversity is managed, the role diversity plays in developing decisions and finally the role of diversity in developing programs are part of everyday discussions. All staff recognize the importance this issue plays in accomplishing our mission."

The benefits of valuing diversity go far beyond greater workplace harmony. A monocultural workforce often feels stagnant and inbred, with the same ideas and attitudes circulating over and over. That staleness is blown away by the fresh air of diversity. As one chief officer said, "Diversity is like getting a transfusion. Thank God for a new approach, something to chew on instead of Cream of Wheat every day!"

Public relations improve when a fire department can show the communities it serves that it values diversity. This enhances the creditability and effectiveness of fire department outreach programs, from firefighter recruitment to community-group liaisons, fire safety education, and arson-watch programs. When firefighters of all races and both sexes, on and off duty, speak well of the department and respectfully of each other, the fire department's public image is enhanced. Every time a piece of apparatus goes out the door or a member of the public visits a fire station, the fire department advertises its support for diversity. The department then can market a diverse emergency service to its communities effectively.

Leadership: "walking the walk"

Moving towards a workplace where cultural diversity is valued means challenging the comfort zones of people entrenched in the status quo. Those who promote change within an institution thus face strong reactions. The X's who represent change often become the convenient targets of that reaction. Fire service leaders must ensure that black firefighters, women firefighters, etc., do not shoulder the entire burden of fire service cultural transition. Deliberate change in an organization happens from the top down. If a fire department is to move from the melting pot to the mosaic, the chief and top management must take the initiative to change their own attitudes and behavior first in order to redesign policy and provide the education that will implement change throughout the department.

For top management to be able to support diversity effectively, its members must understand and be able to identify the cultural differences that exist within the workforce. Managers should be aware of their own stereotypes and assumptions, and learn to listen to people positively, not discounting ideas they don't agree with or that come from someone with a different background. Managers should promote diversity education actively and see that all employees have access to the information, people, and other resources they need to do their jobs. A good manager of a diverse workforce will encourage constructive communication about differences instead of pretending everyone is alike, and will treat people in the workforce with fairness instead of with a cookie-cutter uniformity.

Valuing diversity must become a permanent part of the way the fire service does business. Institutionalizing changes will help make sure they survive a change of chief, future budget cuts, or a new political climate. Your department's written policies and procedures should reflect its commitment to diversity. For example, some cities work against retaliation by having a system that automatically red-flags the name of anyone who files a complaint or grievance and is subsequently denied a promotion or special work assignment. The name goes directly to a city agency outside the fire department, such as EEO or the city council, to be reviewed without the individual having to file a complaint.

What is the future for a fire service leader who chooses not to implement these changes? The chief of a diverse workforce who does not deal positively with diversity will find his or her leadership weakened. A chief who "talks the talk" about valuing diversity but fails to support its implementation loses credibility and ensures the program's failure.

Weak leadership and a superficial commitment to diversity can mean reduced respect for the fire department in the community by city government. This, in turn, can translate into reduced financial and political support, affecting how much clout the chief or department has in other areas such as a fight against EMS privatization or to get a new fire station. The fire department's autonomy may also be threatened, inviting micromanagement from its governing body over issues such as hiring or promotions.

Cultural diversity training

Cultural diversity training, or diversity awareness training, helps move the fire department away from a monocultural philosophy towards one that respects and supports cultural differences. It refers to personnel training that educates members of an organization about prejudices, stereotyping, and the positive aspects of workforce diversity.

The training can be focused in various ways, depending on the needs of the department. It is important for the fire chief to be aware of these options, and of other types of helpful training that are available. Some programs, such as antiracism or antihomophobia training, focus specifically on issues of race or sexuality. Others provide education on sexual harassment and other gender-linked issues. Auxiliary training on conflict resolution, communications skills, and mentoring will supplement diversity training. Some of these types of training should be done before diversity issues are addressed, in order for the diversity training to be more effective.

Just like training on firefighting skills, cultural diversity training is not a one-shot deal. It is an ongoing process of education and reinforcement a fire department undertakes to help the workforce be more effective, harmonious, and productive. The fire chief should assess the needs of the department carefully in order to design a long-range training program that begins with intensive training for the management staff.

Volunteer departments and diversity training

Volunteer fire departments can benefit greatly from cultural diversity training. Disharmony within volunteer ranks can lead very easily to firefighters or officers leaving the department, a loss most volunteer organizations cannot afford. Providing training that will help bridge cultural gaps and other interpersonal barriers can mean the difference between a smoothly running organization and an inefficient, unreliable one. It also can increase the possibility of recruiting new members from among previously excluded groups in the community. Because volunteer departments are so much an integral part of their communities, a fire department that reflects the community's diversity will be much more in touch with its needs.

Financial constraints will call for creative solutions to make cultural diversity training possible for volunteer departments. Consultants' donated time, joint training with other departments, and grants or other funds from community sources or State fire agencies, all may be options.

Several aspects of the fire service breed and reinforce prejudice on the part of its dominant group:

- the dominant group's unfamiliarity with other cultures;
- the dominant group's perception of its ways as "right," and not as just one culture out of many;
- the dominant group's perception that its dominance is threatened; and
- a system of direct competition with coworkers for limited rewards such as promotions and training opportunities.

Other factors offer a foothold for developing acceptance and a pluralistic outlook. Many of the factors that strengthen the bonds among white male firefighters can extend to minority and women firefighters as well, such as:

- a mutual purpose on the job;
- frequent contact with each other in the pursuit of common goals;
- the sharing of intense experiences; and
- interdependence of crew members in critical situations.

Good cultural diversity training for the fire service identifies, discusses, and limits the effect of forces, such as those in the first group, that encourage intolerance. It then identifies positive and cohesive forces, such as those listed in the second group, and enlists them to combat the destructive ones. For example, most firefighters know what it feels like to be picked on because they don't fit the norm. This experience can be explored to build solidarity and encourage firefighters to be aware of, and end, some of the behaviors that cause pain to others.

Diversity training leads to a view of the workforce as multicultural, rather than of the women and minority men as "different." This relieves some of the pressure on X's to fit in at all costs, deny their identities, and try to become just like everyone else. Learning to value diversity means discovering that no group's ways are "right" all the time or for everyone, and that one's own culture has value in the workplace equally with all others. It means understanding how one's own experiences influence and limit one's comfort level around people who are different from oneself. Most importantly, it lets employees gain an awareness of how they fit into the workforce mosaic.

Why do diversity training?

Miscommunication between men and women in the workplace has a strong cultural component. The more forcefully male culture expresses itself in a given fire department, the more difficult it will be for many women to find productive and happy careers there. Understanding differences between women and men as cultural differences lifts the burden off individuals; training on these issues gives all firefighters the tools to create a workplace that works for everyone. These tools help all employees through the stresses and transitions that accompany workforce change.

Lawsuits over on-the-job discrimination and sexual harassment can be expensive, with awards often in the $250,000 to $500,000 range. But the primary reason to do cultural diversity training is not to stay out of court. Fire chiefs provide cultural diversity training for themselves and their workforce because it's part of good management. Even if things never go as far as a lawsuit, discrimination and disharmony are costly in terms of factors that cannot be assigned a dollar value: poor morale, loss of employees from groups the department has been working to recruit and retain, and a discredited reputation for the department within the community.

Cultural diversity training is not about "staying out of trouble." It is about managing change progressively, and taking full advantage of the wide range of resources your firefighters and officers bring to the job. Some of the benefits of workforce education in diversity issues are

- A more positive work environment.

- Enhanced trust, respect, and unity of purpose within the workforce.

- Improved productivity: employees can focus on the job at hand, not on protecting themselves against real or perceived threats. No one is left out of the system or forced to act like someone they're not.

- Enhanced group problem-solving abilities, as a wider variety of ideas and viewpoints is supported.

- Reduced stress, resulting in fewer injuries and reduced use of sick leave.

- Greater understanding of and respect for different cultures, resulting in an improved level of service to the community.

City-wide diversity training

An important vehicle for success in the 1990's is an understanding of the value of diversity and of how to tap the potential of a diverse workforce. Developing this understanding is the core of the work done by the Centre for Organization Effectiveness.

The City of San Diego has used the Centre to help improve its commitment to diversity since 1991. Since that time, more than a thousand employees have attended four-day educational sessions, which provide a safe environment for learning and dialogue. Discussions focus on prejudice, systemic oppression, differing perspectives and added value brought by diverse groups, sexual harassment versus mutual attraction, men's and women's style differences, conflict resolution, clear communications, and action planning. Many qualitative changes have taken place as a result of the training, in the way business is done, people are treated and policies are carried out. Many city departments and divisions have created their own diversity committees. Through improved communication and departmental participation, more EEO issues are being handled internally and fewer formal complaints are being filed.

The Centre receives requests for assistance in forming Diversity Commitment programs from public, private and nonprofit agencies. It also can help organizations implement diversity initiatives to analyze their organizational structure and identify how this structure affects workplace inclusiveness. The Centre trains management and supervisory personnel in managing inclusion, building high-performing work teams, and tapping an organization's full potential.

The Centre for Organization Effectiveness can be contacted at 1250 Sixth Ave., Suite 150, San Diego, CA 92101; 619/685-1340.

Teambuilding: Training on communications and cultural diversity gives firefighters ways to talk about things they often feel they can't talk about. "I don't know how to ask her that without maybe offending her..." "It might be discrimination if I say that..." The knowledge that stems from diversity training creates a shared language in which to discuss problems. Better understanding improves communication, communication builds trust, and trust enhances teamwork.

Ending discrimination: Knowledge about cultural issues helps supervisors and coworkers intervene effectively when conflicts occur in the workplace. Improved communication across cultural barriers helps individuals work out misperceptions that otherwise might lead to a discrimination complaint, and helps stop inappropriate behavior before it becomes a major problem. Diversity training allows everyone to identify and work through their own stereotypes about others. It also lets everyone take pride in their own background, making it less likely anyone will feel compelled to put up with harassment and unfair behavior.

Evaluations: With an understanding of communications style differences, and of one's own stereotypes and preconceived notions, supervisors can evaluate all employees' performance more accurately. A recruit firefighter who is quiet and unassertive, or the one who asks lots of questions, isn't necessarily less competent than the one who is the first to do everything and never asks for clarification. A firefighter who doesn't look the way the officer has always thought firefighters should look still may be a good firefighter. Strengths and weaknesses can be assessed more clearly when the officer has an understanding of the firefighter's background and of his or her own expectations.

Workforce diversity: Continued success in recruiting, retaining, and promoting employees from outside the dominant group is unlikely if the workplace does not value diversity. Conversely, in a workplace where differences are viewed as assets, people from a wider range of background are much more likely to feel comfortable, develop a sincere commitment to the job, express different viewpoints or new ideas, and excel.

A State source of diversity training

The Virginia Fire Services Board and Department of Fire Programs have held an annual Equal Employment Opportunity/Affirmative Action Symposium since 1988. The symposium offers firefighters and officers an exposure to the basic concepts of cultural diversity, with the expectation that this opportunity for learning and increased awareness will have a positive impact on both attitude and behavior at work. Participants report favorably on both the content and impact of the symposia presentations: nearly 75 percent indicated that their attitudes towards minority and women firefighters had changed as a result of their participation in the event.

Information about the Symposium can be obtained from the Virginia Fire Services Board, Department of Fire Programs, Parham/64 Building, Suite 200; 2807 Parham Road, Richmond, VA 23294; phone 804/527-4236.

Delivering the training

Some fire departments handle antiharassment and cultural diversity training in-house. Because hiring outside professionals can be expensive, the temptation to "just have Captain Smith take everyone over the harassment policy" often wins out. But this is not diversity training. Managers should consider the many disadvantages to such short-term cost cutting carefully. Most fire departments, fire districts, municipalities, and counties do not employ people with the expertise required to deliver effective cultural diversity training.

Trainers must be knowledgeable in cultural diversity issues, and also must have good counseling skills to handle verbal conflicts and situations charged with strong emotions. Trainers should either be skilled facilitators, or a separate facilitator should be involved in all sessions. Without these skills, the training often unleashes a free-for-all discussion of sensitive and controversial issues without being able to channel the energy positively or resolve the conflicts and accusations that arise. Minority-group members in such sessions often feel ignored, threatened, or criticized. White men tend to dominate the discussion yet still feel they're not being heard or their ideas are not being validated; they often feel collectively blamed for things they as individuals never have done. When the door is opened to people's deeper feelings, strict control and direction must be provided in order to keep the space safe. The training is seen as a license to talk about normally taboo subjects; ground rules for the discussion should allow everyone to raise his/her their concerns but make it clear that "beating up" on any one person or group will not be permitted; trainers must be able to enforce those rules tactfully but effectively. If they cannot, the resulting damage may be difficult to repair.

One union's support of women

Since 1985, the San Diego Fire Fighters (IAFF Local 145) have sponsored and supported a Women's Issues Committee. The committee exists to allow the union to have input from women firefighters about their concerns over issues such as parental leave, recruitment, uniforms, and station facilities. While the committee was created and is funded by the local's executive board, decisions as to how its members would be selected and how the committee would operate have been left up to its chair.

The committee meets monthly at the union office and has full access to its facilities. A liaison from the executive board also attends the meetings. The committee has been involved in projects such as the establishment of a maternity leave policy, development of a recruitment packet funded by the local, improvements in the fit of women firefighters' uniforms, and creation of a role-model program for new recruits.

The impact of the committee has been both to make the local more responsive to women's needs, and to make women more interested in being involved with the local. As a result of the committee's work and the local's support, the name of the "California State Firemen's Association" was changed in 1990 to the "California State Firefighters' Association." CSFA has also established its own women's issues committee that serves as a resource to all California fire departments.

The Women's Issues Committee of Local 145 is a creative response to gender integration concerns. Other IAFF locals have, with guidance from the International, established Human Relations Committees that serve a comparable purpose with respect to ethnic minorities as well.

(Thanks to members of Local 145 and its Women's Issues Committee for the information presented here.)

Inexperienced or unskilled trainers simply may read through the department's applicable policies, warn attendees of the penalties for violations, and ask if there are any questions. This is of limited value. Effective diversity training is participatory and interactive. Trainers must be able to develop and run realistic scenarios, role-playing and other exercises to involve participants in an active and challenging learning process.

Inexperienced trainers also may expect women and people of color in the group to become an instructional part of the training: to let their personal experiences represent all women, all black people, etc., and to put themselves forward as examples for the benefit of the dominant group's education. The role of all participants as such must be respected by the trainers and by the group. To single anyone out violates the premise of multiculturalism.

Trainers must be able to develop a high level of comfort and trust within the group being trained. Many people find it hard to speak openly about sensitive issues and to discuss workplace problems frankly when the trainer is one of their supervisors. In addition, as will be discussed later, if the training is held as a response to specific problems that have arisen within the department, an in-house trainer who is an officer of the department will be seen as an integral part of the system that allowed the problem to arise in the first place.

Having other municipal or county employees (such as from Personnel/Human Resources, Affirmative Action, or EEO) deliver the training also has shortcomings. In many ways, they combine the disadvantages of outside and in-house trainers. The advantage of hiring outside trainers is that they are skilled professionals who know their material and how to deliver it; the weakness is that they may be perceived as outsiders who don't know "firehouse reality." The main asset of fire department personnel is that they do know firehouse reality; their weakness is that they are not usually skilled diversity trainers. Trainers from other city/county agencies will be viewed with skepticism for being outsiders and also may be perceived as lacking the necessary knowledge and skills. Depending on the past relationship between the fire department and the department providing the trainers, distrust also could be a factor.

If good cultural diversity training is a management priority, it will receive budgetary priority. Hiring a professional consulting firm to develop and deliver cultural diversity training for your department does cost money, but hiring trainers who have both expertise and experience makes it much more likely the resulting program will have a positive result. A cost-effective solution for larger cities would be to hire qualified diversity trainers as full-time employees to do ongoing training for all municipal departments on a rotating basis.

Trainers should be willing to incorporate members of your department into the design and delivery of the program. The trainers should visit the fire stations and department headquarters to seek a wide range of personnel input before the content of the course is finalized. This helps them get a better understanding of fire station life, further demonstrates management's commitment to the program, and ensures the best possible fit between your needs and the program's content.

The program should be designed specifically for your department, reflecting its current level of acceptance of diversity and addressing employees' primary issues and concerns. Its schedule should accommodate firefighters' shifts, the geographic distribution of employees throughout the city or district, and the demands of emergency response. Crews from each station should get information about the diversity of people living and working in their response district as well as in the workforce, to help convert general concepts into useful information.

For members of fire departments that do not provide cultural diversity training, other sources often are available. State fire academies may provide or be willing to develop programs that can be brought to individual departments or offered on a regional basis. Universities with governmental affairs departments sometimes will work with fire service agencies to develop and deliver programs. Many fire service conferences now offer workshops and speakers on cultural diversity issues and resources, as do some state agencies: for example, the Commonwealth of Virginia holds an annual two-day EEO/AA symposium for the fire service. (*See sidebar, page 99.*)

While all personnel must receive the training, education and the commitment to change start at the top. This means the chief and senior staff will go through an intensive training program before the firefighters' training begins. All of the department's officers should receive specially designed training that addresses their needs and roles as supervisors. Officers and chiefs also should be present at each training session for firefighters, not to supervise or to control behavior, but to listen and learn, and to demonstrate by their presence management's commitment to the program. In order to make a free discussion possible, part of the training also may involve peer group sessions, where officers are not part of the firefighters' groups.

Training should be held on a regular basis, following an overall plan. Later sessions should build on earlier training to improve understanding and communications. The trainers should have an evaluation process to gather feedback, and make changes to the training based on this information. Training on cultural diversity issues also should be included in the recruit curriculum or other basic firefighter training, and in all officer development courses run by the department.

Pitfalls to avoid in diversity training

- Avoid trainers whose styles are not effective with your organization. Examples:

  - Trainers who use an "in-your-face" approach to antiracism or similar work (i.e., "You're racist, you're homophobic, and you'd better deal with it!"). Fire department personnel tend to react negatively to the training rather than taking it as an encouragement to work on their prejudices.

  - Trainers who follow a "laundry list" or "cookbook" approach to diversity (i.e., "1. Don't pat Asians on the head, 2. Don't call African-American men "boy," etc.). This approach feels like "real information," but tends to substitute the memorization of sets of rules for the development of understanding and the ability to work issues out for oneself.

A state firefighter's union and women's networking

After attending the 1995 international conference of Women in the Fire Service, the Secretary-Treasurer of the Michigan State Firefighters Union (MSFFU) was motivated to support women firefighters' networking in his State. To do so, he organized a meeting of career-level Michigan women firefighters, held in Lansing in September of 1995. Approximately 12 women firefighters, as well as MSFFU officials, attended the meeting, which was designed to help more women become involved in their union at the State level, to learn what the MSSFU could do to represent women more effectively, and to discuss isolation and other problems women were experiencing on the job. Travel expenses were covered by the union, and attendees who were scheduled to work on the day of the meeting were allowed to use union

---

- Make sure the trainers you use have the necessary range of knowledge and will deliver the material you want. Some consultants in workforce diversity deal primarily with ethnic issues. Your trainers should be fully familiar with gender issues in male-dominated workplaces and have included those issues as an integral part of their curriculum, not simply tacked on some information about sexual harassment.

- Don't do the right training at the wrong time. Examples:

  - Tackling antiracism work before more basic skills are in place. Fire department personnel should be trained in interpersonal communications and conflict resolution before cultural diversity training is addressed.

  - Holding the training in response to a specific problem that has arisen within the department. When the atmosphere has become polarized and hostile around specific issues, it will be more difficult to do productive training. Management should deal first with the problem, and only have training on the underlying cause or enabling conditions once the initial situation is resolved.

  - Holding training in direct response to the hiring of the first X. This puts that person entirely on the spot, and makes any negative aspects of the training their fault: "We have to sit through this (or spend money on this) because of you."

- Avoid putting too many attendees in a session to cut costs. Group size should be limited to 20 to 25 attendees. Larger group sessions tend to become lectures and do not provide time for each person to speak at any length.

- Don't have unrealistic expectations about what the training can accomplish. Diversity training is a valuable tool, but it is not a magic wand. Not all harassment will stop, nor will all interpersonal problems dissolve, simply because the training has been held.

Other sources of support

Fire departments, municipalities, counties, States, firefighters' union locals, and other fire service groups across the country have developed methods of providing ongoing support to women and people of color on the job. Some of these are described in sidebars to this section of the handbook.

One State's support of women's networking

In November 1995, women firefighters from throughout New Jersey met in a 2-day retreat to attend workshops, network, and recognize their achievements while planning for the future. The event was run by the Rutgers University Center for Government Studies, in conjunction with the New Jersey Division of Fire Safety and the New Jersey Division of Women. One of their goals for the event, which was also backed by Governor Christine Todd Whitman, was to show the State's support for its women firefighters.

In order to publicize the event, the Division of Fire Safety sent out advance information to various fire service publications and conferences. Regional meetings were held in different parts of the State to promote the event and generate interest. As a result, more than 125 of the State's 400 women firefighters attended.

Workshop topics included leadership development, sexual harassment, safety equipment, and survival skills. Discussions also focused on factors limiting the numbers of women firefighters in the State, and ways to eliminate those barriers. A highlight of the weekend was a surprise visit by Augusta Chasan, 93, who in 1983 broke tradition by becoming an active member of the Roosevelt (NJ) Volunteer Fire Company. Ms. Chasan received a standing ovation and was later surrounded by women wanting to have their picture taken with her.

One of the most significant outgrowths of the retreat was a new networking organization, Fire Service Women of New Jersey (FSWNJ). Meetings of the FSWNJ are held bimonthly, and there are regional meetings as well. FSWNJ has a column in the newsletter of the Division of Fire Safety and expects to have its own newsletter in the future. A second statewide conference is planned, and FSWNJ and the Division of Fire Safety hope to increase the numbers of women and men attending.

For more information about this networking effort, contact State Executive Liaison Marylain Kemp at 609/633-6106.

Mentoring programs. These match up an incumbent firefighter with a new firefighter, usually a new recruit. The mentor provides personal contact, information on unofficial "rules" and behavior standards within the organization, the benefit of the mentor's experience as guidance for the younger firefighter, and, if need be, a sympathetic voice or a shoulder to cry on. Programs are voluntary and, though facilitated by management, operate at the individual level to provide crucial support for the firefighter early in her/his career or volunteer service.

Workforce diversity steering committees, Women's Issues Committees, Human Relations Committees (functions of department or union). These committees or task forces can operate within a fire department or union local, or on a city-wide or county-wide basis. Their function is to provide information to management and/or union leadership regarding the concerns of women and minorities on the job. (*See sidebar, p.* 100)

Local networks and national organizations. Women firefighters have created local support networks in many parts of the country. These range from informal groups that give a few women the opportunity to share problems and solutions over breakfast once a month, or more established organizations that hold regular meetings, put out a newsletter, and offer workshops and speakers on topics of interest. Their networking, support, and problem-solving functions can be critical to women's performance and longevity on the job. One local network of African-American women firefighters, for example, has developed a program that helps women prepare for fire departmental jobs, from strength training on through the application and interview process. Fire departments should be involved with national organizations and support their members--of all races and both genders--who want to participate in events of those groups.

Other Aspects of Diversity

African-American women firefighters

Women firefighters of all races face many barriers when entering a field such as the fire service. For women of color, the obstacles are even greater. African-American women have been career firefighters for more than 20 years, facing the dual challenges of racism and sexism.

A 1995 study of women firefighters found "persistent and pervasive patterns of exclusion" affecting black women. This exclusion took many forms: inadequate training, close supervision, open hostility and silence from coworkers, stereotyping, and lack of support.

One woman in the study had been required to chop down a live tree as a purported test of her ability to use an axe. Another "asked a senior male colleague for help with a leaking air pack (but) received no constructive instruction and was subsequently written up for presumed negligence, and was referred for additional training."

Racism and sexism interact in ways often detrimental to African-American women. In many cases, women of any race who had won acceptance on the job distanced themselves from black women who had not been so fortunate. In many cases, restroom facilities were improved, uniform fit addressed, and sexual harassment training held only after the first white women firefighters were hired, although black women were already on the job. While 76 percent of white women felt they were treated differently and negatively because they were women, 92 percent of African-American women felt this way.

Relationships with African-American male firefighters were mixed for black women. Most had some supportive relationships, but many also had encountered black men who took the opportunity of women's presence on the job to unify with white men in opposition to the women, thus relegating African-American women firefighters to a new, lower place in the pecking order.

In the same way, many white women firefighters chose to bond with white men at the expense of the black women, allowing racism and a precarious need for acceptance to divide the department's women firefighters.

A clear conclusion of the study is that race and gender are "omnirelevant and inseparable" for the African-American woman firefighter. In every aspect of her work life, she is viewed in terms of both her race and sex. Despite these burdens, African-American women firefighters have succeeded on the job, excelling as firefighters and advancing through the ranks in fire departments all across the country.

Source: Yoder, Janice D., and Patricia Aniakudo, "'Outsider Within' the Firehouse: The Impact of Race and Gender on the Social Interactions of Black Women Firefighters." University of Wisconsin--Milwaukee, 1995.

An Asian perspective

Growing up female and Asian-American, I never thought about becoming a firefighter. People in Asian cultures generally raise their children to strive for white-collar careers, not blue-collar jobs. Additionally, firefighting was and still is considered man's work. The religious and patriarchal influences dominant in Asian cultures reflect this prevailing viewpoint.

In many ways, I was raised like a typical Asian woman. In my family, honor, respect and obedience were stressed. While I was not told what profession to enter, it was understood that I would attend college and use my educational background for an appropriate career. After finishing college, however, I discovered the fire service. My parents' reaction to my decision to become a firefighter made it clear that this was not a proper choice for an educated Asian woman. As the years have passed, however, my parents' acceptance of, and even pride in, my being in this profession have grown.

My choice proved to be the right one for me, because in other ways as well I am not a stereotypical Asian woman. I relish working in a field that has historically been for men, particularly white men. I express myself loudly sometimes, if need be, on the fireground, and I am not the quiet, Asian person I am expected to be.

That is not to say that the fire service has been without its challenges for me. On the contrary, like most Asian people, I am small in stature, which presents a great challenge for a firefighter. I've overcome difficulties regarding the fit of my personal protective equipment as well as weight/strength/height ratios of equipment, apparatus and required evolutions.

Overcoming challenges is part of being a good firefighter. Most importantly for an Asian woman, being a good firefighter is determined by more than one's sex, size, or culture.

Source: Yamane, Grace; "An Asian Perspective." *Women in the Fire Service Quarterly*, Winter 1997, p. 2.

Antigay prejudice: a management issue

Prejudice against gay people--homophobia--is a problem for fire service managers in the same way that racism and sexism are. Like other forms of bias, it interferes with the smooth functioning of the workforce. It impairs the job performance of gay and lesbian firefighters and officers, as well as that of employees who have gay family members and must endure derogatory jokes and comments at work about the people they love. It can also affect the quality of service provided to gay men and lesbians in the community.

Indirectly, homophobia feeds sex discrimination and sexual harassment. Women applicants have been denied jobs simply because someone thought a strong, assertive woman "looked like" a lesbian. Women on the job have been coerced into putting up with harassment and even sexual assault by the threat of being called a lesbian if they object. The fear of being labeled in a way that can cost one one's job, one's personal safety, and the affection of friends and family is a powerful motivator. This same fear can prevent women from getting together to discuss and resolve job-related problems, even in informal settings or department-sponsored groups.

In many places, equal treatment of lesbians and gay men in the workplace is simply the law. A large U.S. police department found guilty of harassing its gay employees had to pay more than $750,000 as a result, and was placed under court order to recruit lesbians and gay men as police officers and firefighters, to screen all city applicants for anti-gay attitudes, and to discipline or terminate supervisors who exhibit antigay behavior. While the threat of losing a lawsuit should not be the main reason to work for fair treatment in the workplace, such cases serve to show what can happen when these issues are disregarded.

What should fire chiefs do?

Review your department's hiring and promotional practices to see that they do not discriminate against lesbians or gay men, intentionally or unintentionally. Many fire departments still have questions on their polygraph tests about applicants' sex lives; these may be of questionable legality.[1] Implement policies that assign and promote people based on their qualifications, to minimize the impact of all forms of prejudice on employees' career opportunities.

Extend health insurance coverage to partners of gay and lesbian firefighters in the same way that it is offered to firefighters' spouses. Examine the impact of other benefits, such as sick leave, parental leave, and death benefits, for their impact on, and accessibility to, gay and lesbian employees and their partners.

Include sexuality in your department's statement of nondiscrimination, and enforce this policy. Ban anti-gay language, jokes, cartoons, etc., in the same way that racist and sexist language and materials are banned from fire stations and all department activities.

Include homophobia issues in cultural diversity and antiharassment training. Education will help make it clear to department members that sexuality is about one's identity, and that being gay or lesbian has no more to do with sexual activity that being heterosexual does.

Working against homophobia in the fire service means raising controversial issues that many people feel strongly about. Stopping discrimination against gays and lesbians on the job is not an endorsement of sexual activity of any kind: it is a step towards fairness and equality. Tackling these issues aggressively, with sensitivity, and with a sincere commitment to fairness, is the most effective way to stop the damage this form of prejudice can do to your fire department.

[1] *Woodland v. Houston*, 1995 U.S. Dist. LEXIS 2749 (S.D. Tex.)

Peer mediation: a solution that works

Somewhere between bearing in silent pain the injuries inflicted by a fellow firefighter and taking on the stress of filing an EEO complaint lies a middle course. Many fire departments, seeking to provide their personnel with an informal forum for resolving interpersonal issues, have come up with an idea that is effective, innovative and user-friendly. That idea is peer mediation.

Being labeled as unassertive if they opt to endure a bad situation, and a troublemaker if they consult with an EEO officer, women and minorities often find themselves caught in a bind. Peer mediation programs offer a partial solution. While such programs do not replace the EEO officer or take away an employee's right to file a discrimination complaint, they do provide an alternative to these options.

Fire department peer mediation programs began in the San Francisco Fire Department in 1988 in response to a mandate of a consent decree over hiring practices. The decree required the department to "establish and maintain an informal mediation process to handle internal disputes."

Under the program thus developed, volunteers from both the uniformed and civilian staff undergo an intensive training course: 7 full days over a 2-week period. Students spend the first 2 days learning the elements of communication, the third exploring sources of cultural conflict in the workplace, and the final four learning and practicing mediation skills and techniques.

Mediators are used in both formal and informal ways. Informally, they use their training to defuse conflicts in their station, to communicate better with coworkers, and to manage potential conflicts encountered during emergency responses. Many also report being able to use the techniques off the job, such as in situations at home.

The formal process begins when two parties to a conflict on the job agree to have their dispute mediated, and select a mediator from the department's list. The process is confidential and voluntary, and either person may withdraw at any time. Following is the four-step approach used by San Francisco, which is typical of programs in use throughout the country.

1. Each person tells the mediator her or his version of the situation. The ground rules require mutual respect (no interruptions or name-calling) throughout the process. The mediator asks questions to clarify the issues and each party's feelings about it.

2. The mediator opens communication between the parties, helping each understand how the other views the conflict and is affected by it. At this point, the parties communicate directly with each other about specific issues.

3. The mediator and the parties discuss options for resolving the conflict, both for the present and in case it should arise again.

4. The two parties jointly agree on a resolution.

The SFFD program was developed by a team of consultants and in a fire department task force. When it was first established, 90 department members volunteered for the mediator training. As word of the program's effectiveness spread, the list grew. A small group of mediators has now gone through an advanced course and provides in-service training for the department and continuing education for the mediators.

In Prince George's County, Maryland, two task forces worked during 1993 and 1994 to develop ways to support, strengthen, and better serve the arena of equal employment opportunity. The Gender Issues Task Force and the Cultural Diversity Task Force came up with the same suggestion: a peer mediation program.

Training was provided by a private company that had done peer mediation training for the county's public schools. The fire department mediation program was designed to serve two functions: to mediate conflicts arising between employees, and to provide an informal and confidential advisory service on EEO issues where employees could ask questions and vent frustrations without setting the official EEO machinery in motion.

For the first year and a half of the program, the latter function was used far more often than the former. "Peer mediation programs develop slowly, much like the development of Critical Incident Stress Debriefing in the early 1980's," explained Captain Elizabeth Jackson Redding. "CISD is now widely used and acknowledged, but it evolved gradually, and only because of the hard work, training and dedication of the program's advocates."

Prince George's County was fortunate in having Captain Redding on its staff, as she is also a licensed Employee Assistance Program (EAP) counselor carrying malpractice insurance. The Peer Mediation Program thus could be set up under her guidance and license, much as a paramedic acts under the license of a medical director. This provided peace of mind for the mediators and an essential promise of absolute confidentiality to the firefighters in the field.

The Chesterfield (Virginia) Fire Department's Human Relations Committee came up with peer mediation as an alternative to the grievance process. "We are hoping to catch potential problems while still small enough to solve on an informal basis," said Deputy Chief F. Wesley Dolezal. Captain Redding will share her expertise with Chesterfield as their program gets off the ground.

The jurisdictions that have experimented with peer mediation find it works. The idea is spreading as progressive fire departments look for ways to prevent and resolve employee conflicts before they escalate into major problems. Peer mediation can be a positive step towards reducing the apprehension and suspicion sometimes associated with discrimination issues in the workplace.

Firefighter protective gear

According to data from Women in the Fire Service's 1995 survey of fire service women, women firefighters continue to have difficulty getting protective gear and clothing that fit properly. While manufacturers over the past five years have seemed to be responding to the increasing demand for women's fire gear, comparisons with data from the Women in the Fire Service's 1990 survey show the problem, surprisingly, is getting worse instead of better.

Five years ago, 51 percent of women firefighters had problems with gear fit. In 1995, 58 percent--287 out of the 495 women in suppression roles responding to the survey--were functioning in their jobs with one or more items of protective gear that did not fit. Of the 42 percent who said all their fire gear fit, some added comments such as:

"Took over eight years to get a SCBA facepiece to fit."

"Have been fighting for women's uniforms for fifteen years!"

Several women noted they had finally had to buy some or all of their own gear in order to get it to fit.

Following is a list of the items causing the most problems, the percentage of women experiencing problems with each item, and some of the women's comments about the gear and their attempts to get a better fit. All percentages have increased since 1990 except for two categories: the fit of turnout/bunker coats showed a moderate improvement, and the percentage of women whose bunker pants did not fit stayed the same.

Firefighting gloves: 31 percent of women firefighters said theirs did not fit (as compared to 25 percent in the 1990 survey). Forty-three percent of these women said their gloves were too big; 13 percent that the fingers were too long. Ten percent commented that the gloves were too bulky or thick, and that it was difficult to grip things or otherwise work in them.

Rubber firefighting boots: 19 percent (13 percent in 1990)
Forty percent of women whose boots didn't fit said their boots were too big. Several women resorted to wearing extra socks to fill up the extra room in their boots: one said she wore five pairs. Thirteen percent said their boots were too wide in the foot or the heel. One woman who had been a career firefighter for 10 years said of her boots: "They do not fit in length or width. Very uncomfortable; fall off in fires if not held on by bent feet." Several women noted that boots in half sizes would be helpful; several others commented that if the boots were small enough to fit in the foot, they were too tight in the calves, or too short. One woman noted a particular safety problem for apparatus drivers: "While driving (using clutch), top of boot gets caught under the seat."

Turnout/bunker coat: 16 percent (down from 19 percent in 1990)
More than a quarter of the women whose coats didn't fit (27 percent) said their coat was too long; 20 percent said it was too big. Another 27 percent said their coat was too narrow in the hips. Women whose coats did fit through the hips often reported it was then too big in the chest, shoulders, or sleeves. Several women noted that their coats didn't fit even though they had been custom made for them. In all, 20 percent of the women responding to the survey said they could not hook the lowest hook on their turnout coat, either at all or when wearing turnout pants. (This number includes some who indicated they had no fit problems with their protective gear.)

Turnout/Bunker pants: 14 percent (also 14 percent in 1990)
Thirty percent of the women whose turnout pants didn't fit said the pants were too tight in hips and thighs, or too big in the waist. Nineteen percent said their pants were too big. "Two of me can fit in them," one woman said. Fourteen percent said their pants were too long, either in overall length or in rise (the

measurement from the waist to the top of the inseam). As with the turnout coats, several women noted that their pants had been custom made and still did not fit.

SCBA facepiece: 14 percent (11 percent in 1990)
More than a third (34 percent) of the 14 percent said their facepiece was too big, or leaked at the temples. A few noted that they had been forced to use an uncomfortably small mask because they could not get a proper seal with the next larger size. Several complained about lack of size options from some manufacturers. Others found the proportions of the mask wrong for them, particularly that they needed a longer, narrower facepiece. Fourteen percent noted that their facepiece did not work well when their face was sweaty, or that the facepiece slipped when in use.

Helmet: 13 percent (11 percent in 1990)
Half of these women said their helmet was too big. This was true of helmets with ratchet-type adjustments as well as those without: "Have it as small as possible; still falls off." Other women found the helmet's interior shape didn't fit their head: "I have a very small head, and the band becomes too round when I ratchet it down to fit. Helmet is constantly slipping."

SCBA harness/pack: 9 percent (not asked on 1990 survey)
Straps in general were a problem for 20 percent of women whose SCBA's did not fit. Several women commented that the straps on one particular manufacturer's SCBA harness tended to loosen rather than staying adjusted. Other complaints included shoulder straps that were too stiff and tended to come off, or that restricted arm movement. A number of women found the straps too long: "(They) either hang down and catch on things, or I have to tuck them in, which makes quick removal of SCBA hard." Sixteen percent found the chest straps uncomfortable or poorly designed, particularly when a PASS device was worn. Twelve percent said the waist strap would not adjust tight enough. Frame design was a problem for another 20 percent, most commonly that the frame was too long.

Nomex/PBI hood: 2 percent (not asked in 1990 survey)
Even a piece of protective clothing as simple to design as a hood does not always fit. Women found their hoods too loose around the face, too big, or not long enough.

General comments about gear fit and design:

"All gear/clothing made to fit male body, not female."

"Would like weight of turnout gear on hips instead of shoulders."

"With Federal wildland equipment, it's not a gender problem. The gear isn't right for anyone!"

Gear interface problems:

Nearly a third of the women responding to the survey (31 percent) reported that their SCBA knocked their helmet forward or off. One noted that this was a problem for male firefighters on her department as well, and a few felt it was a function of the long rear brim of specific styles of helmet. The problem, however, exists with all styles, and has much to do with how the SCBA frame and bottle fit a shorter firefighter's body.

Many women wrote in other gear interface problems they had experienced. Several women noted that they couldn't use the pockets of their turnout coat or pants when wearing an SCBA. The helmet/SCBA interface was a problem not only regarding the bottle, but also the facepiece: "SCBA mask pushes helmet back on head; helmet not very secure over it." The reverse was true for one woman: "Facepiece seal compromised when helmet is on."

Another point of interface that caused problems was at the wrist. "Coat wrist cuff loses elasticity and doesn't allow glove/cuff overlap." And the reverse: "Long coat sleeves push gloves off."

Responses:

Seventy-six percent of the women responding to the survey said they had advised their fire department of current or past problems with gear fit. Twelve percent said they had not advised the department.

Twenty-three percent of women who advised their department of gear fit problems said the department took care of the problem(s) promptly. Even for those who checked this response, fit problems sometimes remained: either one problem was fixed but not others, or the situation was only improved and not resolved entirely.

Thirty-one percent said the department eventually took care of the problem. "Eventually" was clarified in some cases: "Fixed when a friend took over turnout ordering." "It took them over ten years."

Twelve percent said the department made a good-faith effort but got no results.

"Bought from manufacturer advertised as selling women's gear--but they only added cinch strap to bunker pants fitted for men."

Eighteen percent said the department made a token effort at resolving the problem, with no results. 13 percent said the department had refused to order gear that might fit from a different manufacturer. "City contract bidding requires he who is cheapest wins," one woman pointed out. Another 13 percent were told by their department it would cost too much to take care of the problem: "Said they had made a 'good deal' with the company on men's turnouts, so no women's were ordered." Nineteen percent of the women said their department was still working on the problem.

"I was fitted for new gear two years ago. Gear came in and was still too large. Working on ordering a different type from a different company.

"'Still working on it' will go on forever."

Comments about efforts to get better-fitting gear

"I waited two years (before bringing up the issue) because I didn't want to seem like a complainer."

"The problem with my facepiece was totally ignored for years, despite efforts on my part to get information about options. The problem with the gloves were eventually addressed, but the gloves still don't fit. The department is more responsive to complaints about gear fit than it used to be, but still addresses such problems at their convenience."

"Their response was to give me new gloves with the same problem as my old ones."

"It took 18 months to get my turnout gear. The sleeves were about 8 inches too short, turnout pants about 4 inches too long, suspenders were unable to get short enough. The pants were way too huge all around. Finally got the fit okay, but the coat doesn't buckle, and I'm still waiting on shorter suspenders--due to the only 'okay' fit of the pants, the suspenders are critical. But at what point does insisting on further adjustment and alteration become whining?"

Conclusions

Much of firefighters' protective gear is still made on men's patterns with no, minimal, or ineffective adjustments for women. Even where gear is manufactured to fit women, there may be no effective availability to the firefighter: difficulties and time delays in getting gear that fits mean women will settle for gear that doesn't. Fire departments do not always make properly fitting protective gear for all personnel a priority; sometimes it is not even a concern.

Nine out of ten women firefighters have had problems with protective gear fit at some point in their career or volunteer service. Seventy-five percent of women firefighters have forwarded a complaint about gear fit issues to their departments. Only half of these had the problem taken care of "promptly" or "eventually," and "eventually" in some cases meant many years.

Sources of protective clothing and uniforms in women's sizes

Key: The company listed is the manufacturer of the product, except as specifically noted. Clothing is made on women's patterns, and footwear on women's lasts, as noted.

"Union" = union-made product
"OSHA" = meets applicable OSHA requirements.
"NFPA" = meets applicable NFPA requirements.
"ANSI" = meets applicable ANSI (footwear) requirements.

Rubber structural firefighting boots

Kaufman: Black Diamond. NFPA. Whole and half sizes 3 to 16, medium and wide.

Morning Pride: Women's lasts. NFPA, OSHA. Whole and half sizes, narrow and regular widths. Smallest size is women's 4.

Neptune: (Imports) Viking #9844. Uses inserts, not women's lasts. Union, NFPA, OSHA. Sizes 5 to 15-1/2.

Ranger: # 31290. Women's lasts. NFPA, OSHA. Sizes 4 to 10, narrow and medium.

Servus: Firebreaker, Firebreaker II, Firefighter. Women's lasts. Union, NFPA, OSHA. Sizes 3 to 13, narrow and wide.

Leather structural firefighting boots

Ranger: #3042, 3044, 3045. No women's lasts. NFPA, OSHA. Whole and half sizes 5 to 13, medium and wide.

Southwest: Eagle. Women's lasts. NFPA, OSHA. Sizes 3-1/2 to 15 A-EEEE; Wildland Firefighting Boots. NFPA, Cal/OSHA, ANSI. Sizes 3-1/2 to 15 A-EEEE.

Warrington: Warrington Pro 2004, 2006, 2008. Unisex military last. NFPA, Cal/OSHA. Whole and half sizes 6 to 14.

Weinbrenner: Thorogood 804-6116. No women's lasts. OSHA, ANSI.

Wildland firefighting boots

Iowa American: (Distributes) Servus. Women's lasts. Cal/OSHA. Full and half sizes.

Southwest: Eagle. NFPA, OSHA, ANSI. Sizes 3-1/2 to 15 A-EEEE.

Warrington: Warrington Pro. Unisex. NFPA, Cal/OSHA. Sizes 6 and up.

Wildland firefighting jackets and pants

Alb: 06, 07, 70, 60. No women's patterns. Cal/OSHA. Sizes Small to XXL.

Fire Gear: Overcoat and overpant. Sizes 28 to 64, plus oversizes. Women's patterns. 1977 NFPA, OSHA, Cal/OSHA.

Lion: [Jackets, pants, and overalls] Nomex and regular. Alpha sized S-XXL. No women's patterns. NFPA, Cal/OSHA.

Safeguard America: 71-J, S-82 MOD, 71-T MOD. Women's patterns. Cal/OSHA. Sizes Small to 6XL in jackets and shirts; pants: waist and inseam in 1" increments as required.

Structural firefighting turnout coats and pants*

Alb: C17080, 0708-0, P270080, 7008-0. No women's patterns. NFPA, OSHA. Custom sizing.

Bristol: NFPA. Custom and women's patterns.

Cairns (Div. of Globe Industries): [High-waisted bunker pants rather than bib-style.] Aegis, Traditional, RS1. Women's patterns. NFPA, OSHA. Sizes XS, S, M, L, XL, XXL.

Fire-Dex: Assault Gear, Series III. Women's patterns. NFPA, OSHA, Cal/OSHA. Coats sizes 32 to 60, pants sizes 28 to 56.

Fire Gear: [30"-length coat only, with pants to interface.] Women's patterns. NFPA, OSHA. Sizes 28 to 64, plus oversizes.

Fyrepel (Div. of Lakeland Industries): Attack, Sterling Heights. NFPA. Sizes XS to 6XL. Individually custom made to fit wearer.

Globe: Globe GX-7, Globe Astra. Women's patterns. NFPA (1971 & 1976), OSHA. Coats sizes 28 to 52; pants sizes 24 to 52.

Iowa American: (Distributes) Quaker. Women's patterns in short-style gear. NFPA, OSHA. Custom sizing.

Lion Apparel: Body-Guard, Janesville. Women's patterns. NFPA, OSHA. Custom sizing.

Morning Pride: [No bib-style bunker pants.] Women's patterns. NFPA, OSHA. Full range of sizes. Dual certified NFPA 1999/1971.

---

*All listed manufacturers and distributors carry bunker coats in both traditional and short styles, and bunker pants in traditional and bib styles, unless otherwise indicated.

Quaker: Traditional style. OSHA, NFPA, Cal/OSHA. Sizes S-XXXXL.

Ramwear: [High-wasted bunker pants rather than bib-style.] 2000, 500. Women's patterns. NFPA, OSHA. Custom sizing.

Firefighting gloves

Alb: 589. NFPA, OSHA. Sizes S to XL.

American: American Firewear. OSHA, NFPA, Cal/OSHA. Sizes XXS to XXL; Cadet sizes S to XL. Unisex patterns.

Fire Dex: 500, 600, 900 series (OSHA); 300, 400 series (NFPA/OSHA). Sizes XS to Jumbo.

Fire Grip: 1000, 200. NFPA, OSHA, Cal/OSHA. Sizes XS to XXL.

Fyrepel: Fyrepel. NFPA. Sizes XS to 6XL. Individually custom made to fit each wearer.

Globe: G-9, GL-4. NFPA (GL-14), OSHA (GL-4). Sizes XS, S, M, L, XL.

Glove Corporation: Wildfire. Cal/OSHA. Sizes S to XXL; also women's size XS. Made on different patterns for men and women. Fireman I, V, VI, VIII; Firefighter. NFPA, OSHA. Sizes XS to XXL.

Knoxville Glove: Fire Guardian, Fire Knox. Union, NFPA, OSHA. Sizes XS to J.

Lion Apparel: (Distributes) Lion G1002, G1004, G1006, G1008. NFPA, OSHA. Sizes S to XL.

Morning Pride: NFPA, OSHA. Sizes XS to J, and made to hand tracings.

Neptune: North Star 2990, 2991, 2993, 2994. OSHA. Sizes XS to XL.

ProTek: NFPA. Sizes XXS to XL.

Shelby: Structural, Firewall 2533, 5221, 5225, 5009, 5011; Wildland 5002. Union, NFPA, OSHA, Cal/OSHA. Women's sizes XS, XXS/7,8; Sizes XS-J; also XXS in non-NFPA version.

Tempo: Tempo Max, Tempo Pro. Union, NFPA, OSHA. Sizes XS to J.

Wells Lamont: Backdraft. NFPA, OSHA, Cal/OSHA.

Hazardous material suits

Abanda: [Level C] Fortress 7100. NFPA, OSHA. Sizes S to 7XL.

Chemfab: [Level A & B] Challenge 6400, Challenge 5000, HazTech Vapor, HazTech Splash. Union, NFPA, OSHA. Level A: one size; Level B: two sizes.

ILC Dover: [Level A] Chemturion. NFPA (Limited use), OSHA. Sizes M, L, XL.

Interspiro: [Level A, B, & C] (Distributes) Trellchem HPS; Trellchem TS, TB, TL; Splash 100, 200, 300, 700. NFPA (Level A only), OSHA. Sizes S to XL.

Kappler: [Level A, B, & C] NFPA. Sizes S to 3XL.

Lakeland: [Level A, B & C] Forcefield, Interceptor, Checkmate, Saranex. NFPA, OSHA. Sizes S to XL.

Interceptor: other sizes available on request.

Lion Apparel: [Level B] Pace Setter. NFPA, OSHA. Sizes XS to XXL.

Trelleborg: [Level A] Trellchem VPS, HPS. NFPA. Sizes S to XXL.

Protective hoods

American: Eagle 10" and 12" headpiece with variety of bib lengths. OSHA, NFPA, Cal/OSHA. Unisex patterns. Hoods available in Nomex, PBI, and Lenzing P84.

Tempo: Uno, Uno III, Uno-L. NFPA. One size.

Firefighting helmets

Bullard: PX, FX, CX, TX Wildfire FH 911. NFPA. Adjustable.

Cairns: Cairns "1010," 970R Eagle, 6602CR Metro, 660C Metro, 990 Intruder, 660 Phoenix, N6A Houston, N5AR New Yorker, 770 Philadelphian, Classic 1000, HP1 Commando. NFPA, OSHA. Universal sizing.

Gallet: F-2000, Fire Brigade, F-2 (wildland). Adjustable depth and circumference.

Iowa American: (Distributes) Safeco. NFPA, OSHA. Adjustable, XS to XL.

Lion Apparel: Lion Revolution. Unisex.

Morning Pride: Lite Force I, II, III, IV, V, VI. NFPA, OSHA. Adjustable; factory assistance for unusual problems.

Phenix: First Due 1500. OSHA, NFPA, Cal/OSHA ANSI. Women's size 6-1/2; Phenix First Due 1500. Women's size 6-1/2.

Rescue Equipment NW: Pacific (NZ) Ltd., R3K/1 Rescue Helmet. Union, ANSI, NFPA (partial). Ratchet adjustable, 21" to 25".

SCBA facepieces

Cairns: Pioneer. NFPA, OSHA. Three sizes.

Draeger: Draeger, Panorama, Nova. NFPA, OSHA. One size.

Interspiro: NFPA, OSHA. One size; small available "in extreme cases."

MSA: Ultravue. NFPA, OSHA. Sizes S, M, L.

Scott: NFPA, OSHA. Sizes (for 4.5 and 2.2) Small, Standard, XL; (for IIa) one size.

Survivair: XL-30, Mark 2, Sigma. Sizes Small and Standard. NFPA, OSHA, MSHA/NIOSH; Sigma Twenty Twenty. Sizes S, M, L. OSHA, NFPA.

Safety shoes

Southwest: [Chukka boots; oxfords] CM-500, OX-600. No women's lasts. ANSI. Sizes 3-1/2 to 15 A-EEEE.

Weinbrenner: Thorogood 504-6100, 504-6000, 504-6265. OSHA, ANSI.

Station uniforms

Artcraft Blazer: [Dress uniforms] S/800, S/801. Women's patterns. Jackets sizes 6 to 24; pants sizes 4 to 26; blazer coat sizes 6 to 42.

Cairns: [Coveralls] Series I, Series II, Series III. Women's patterns. NFPA, OSHA. Sizes XS to XXL.

Fechheimer: [Coveralls, jackets, pants, shirts.] Paramount. Women's patterns (except coveralls). NFPA, OSHA. Pants sizes 4 to 24; shirts sizes 28 to 48.

Iowa American: [Coveralls, jackets, pants.] Quaker. Women's patterns. Jackets only: Union, NFPA, OSHA. Any size.

Lion Apparel: [Coveralls, jackets, pants, shirts] Stationwear; Nomex, Firewear, or Flamex. Women's patterns in pants and shirts. NFPA (shirts and pants), OSHA (all except jackets). Coveralls sizes S-XXXL; women's pants sizes 6 to 24; women's shirts sizes 6 to 20.

Safeguard America: [Coveralls, jackets, pants, shirts] M-6, 74-J, 91-T, S-83, S-82. Women's patterns. NFPA, OSHA. Sizes 32XX-short to 64XX-long; waist/inseam in 1" increments; XXS-6XL.

Topps: [Pants, shirts] Women's patterns. NFPA, OSHA. Squad suits sizes S to XXL regular and tall; pants sizes 4 to 28, shirts sizes 8 to 22. Other sizes available by special order.

Werner Works: [Coveralls, shirts, jackets, pants] Pro-Tuff. Women's patterns. Sizes XS to 3XL; women's pants and coveralls sizes 8 to 22.

Workrite: [Coveralls, jackets, pants, shirts] 1106N, 1104N, 3327F, 4007F, 2404F, 2434F, 2344F. Women's patterns in all except jackets. NFPA, OSHA. Sizes 34 to 54; pants sizes 28 to 50. Maternity uniforms available.

Firefighting proximity suits

Cairns: Aegis Proximity, Traditional Proximity. Women's patterns. NFPA, OSHA. Sizes XS, S, M, L, XL, XXL.

Fire-Dex: Aluminized. Women's patterns being developed. NFPA, OSHA.

FireGear: Women's patterns. NFPA, OSHA. Sizes 28 to 64.

Fyrepel: Fyrepel. NFPA. Sizes XS to 6XL. Individually custom made to fit each wearer.

Globe: Women's patterns. NFPA, OSHA. Coats sizes 28 to 52; pants sizes 24 to 52.

Lion Apparel: Body-Guard, Janesville. No women's patterns. NFPA, OSHA. Coats sizes 34 to 60; pants sizes 30 to 56.

Morning Pride: No bib-style bunker pants. Women's patterns. Full range of sizes. Dual certified NFPA 1999/1976.

Quaker: Traditional style. OSHA, NFPA, Cal/OSHA. Sizes S to XXXXL.

Ramwear: Ramwear. Custom sizing. OSHA, NFPA.

Manufacturers and distributors of firefighter protective clothing

Abanda Protective Apparel
AmSouth Building, 4th Floor
P.O. Box 2028
Decatur, AL 35602

Alb, Inc.
366 Somerville Avenue
Boston, MA 02143
617/666-8111

American Firewear, Inc.
P.O. Box 2064
217 East F Street
Anniston, AL 36202
205/237-5575
800/264-3333
(fax) 205/236-9014

Artcraft Blazers
7502 Thomas St.
Pittsburgh, PA 15208
412/242-0266
800/345-4116
(fax) 412/242-2068

Bristol Uniforms North America Inc.
71 Rosedale Ave. West, Unit C-6
Brampton, Ontario L6X 1K4
905/454-0560
(fax) 905/450-3436

E.D. Bullard Company
1898 Safety Way
Cynthiana, KY 41031
800/827-0423
606/234-6616

Cairns Protective Clothing
(Division of Globe Manufacturing)
Leavitt Road
P.O. Box 125
Pittsfield, NH 03263-0125
603/435-7787
(fax) 602/435-7876

Chemfab/Chemical Fabrics Corporation
Daniel Webster Highway
P.O. Box 1137
Merrimack, NH 03054
603/424-9000
800/451-6101
(fax) 603/424-9012

Fechheimer Brothers
(Paramount Protective Uniforms)
4545 Malsbary Road
Cinncinnati, OH 45242
513/793-5400
800/543-1939
(fax) 513/793-7819

Fire-Dex Fire Clothing
3865 W. 150th St.
Cleveland, OH 44111
216/941-3959
800/241-6563
(fax) 216/941-1130

Fire Gear, Inc.
409 AABC Suite C
Aspen, CO 81611
970/925-2303
800/426-9352
(fax) 970/925-2426

FireGrip
6241 Riverside Dr.
Dublin, OH 43017-5034
800/633-7397
614/798-0013
(fax) 614/798-0015

Fyrepel Products
(Div. of Lakeland Industries)
202 Pride Lane S.W.
Decatur, AL 35601
800/345-7845
205/350-3107
(fax) 205/350-3011

Gallet USA, Inc.
94-B Pleasant St.
Brunswick, ME 04011-2207
207/725-0810
(fax) 207/725-0816

Globe Firefighters Suits
Loudon Road
Pittsfield, NH 03263-0128
603/435-8323
(fax) 800/442-6388

The Glove Corporation
301 N. Harrison
Alexandria, IN 46001
317/724-4481
800/346-4253
(fax) 317/724-9613

ILC Dover, Inc.
P.O. Box 266
Frederica, DE 19946
302/335-3911
800/631-9567
(fax) 302/335-0762

Interspiro, Inc.
31 Business Park Dr.
Branford, CT 06405
203/481-3899
800/468-7788

Iowa American Firefighting Equip. Co., Inc.
P.O. Box 517 Industrial Park
Osceola, IA 50213
515/342-6091
800/342-IOWA

Kappler USA
P.O. Box 218
Guntersville, AL 35976
800/633-2410

Kaufman Footwear
206 Meadow Road
Syracuse, NY 13219
315/468-3952
(fax) 315/468-3952

Knoxville Glove Company
P.O. Box 138
Knoxville, TN 37901-0138
423/573-4555
800/251-9738
(fax) 423/573-4558

Lakeland Industries, Inc.
1 Comac Loop
Ronkonkoma, NY 11779
800/886-8010
516/981-9700
(fax) 516/981-9751

Lion Apparel
6450 Poe Ave., Suite 300
P.O. Box 14576
Dayton, OH 45414
513/898-1949
800/421-2926

Mine Safety Appliance Co. (MSA)
P.O. Box 426
Pittsburgh, PA 15230
412/967-3167
800/MSA-2222

Morning Pride Mfg., Inc.
1 Innovation Ct.
P.O. Box 14616
Dayton, OH 45413-0616
513/454-4925
(fax) 513/264-0075

National Draeger., Inc.
101 Technology Dr.
P.O. Box 120
Pittsburgh, PA 15230
412/787-2207
800/922-5518
(fax) 412/788/5944

Neptune International, Ltd.
6925 216th St., S.W.
Lynnwood, WA 98036
206/776-5399
800/776-5390
(fax) 206/776-7979

Phenix Technology, Inc.
500 Harrington St., Suite B
Corona, CA 91720
909/272-4938
(fax) 909/279-8399

ProTek
852 Sunrich Lane
Encinitas, CA 92024
619/632-0624
(fax) 619/632-0455

Quaker Safety Products Corporation
103 South Main St.
Quakertown, PA 18951-1185
215/536-2991
(fax) 215/538-2164

Ramwear, Inc.
9302 Progress Pkwy.
Mentor, OH 44060
216/639-0137
800/777-9497

Ranger Firefighter Footwear
1914 Cometiz Dr.
Davenport, IA 52802
800/668-6148
(fax) 309/762-0290

Rescue Fire Equipment N.W., Inc.
P.O. Box 72
Redmond, WA 98073
206/836-8515
800/743-0554
(fax) 206/836-8712

Safeguard America, Inc.
P.O. Box 1649
Clanton, Al 35045
205/7550-7710

Scott Aviation
309 W. Crowell St.
Monroe, NC 28112
704/282-8400
800/AIR-PACK

Servus Footwear Co., Inc.
9300 Shelby Road, Suite 300
Louisville, KY 40222
502/327-6100
800/777-9021

Shelby Specialty Gloves
P.O. Box 171814
Memphis, TN 38187-1814
901/360-8982
901/362-9127

Horace Small
P.O. Box 1269
Nashville, TN 37202-1269

Southwest Boot Company
2545 San Fernando Road, Suite 22
Los Angeles, CA 90065
213/223-2465

Survivair
3001 S. Susan St.
Santa Ana, CA 92704
714/545-0410
800/821-7236
(fax) 714/850-0299

Tempo Glove Manufacturing, Inc.
3820 W. Wisconsin Ave.
Milwaukee, WI 53208
414/344-1100
800/558-8529

Topps Manufacturing Co.
P.O. Box 750
Rochester, IN 46975
800/348-2990
800/552-2351 (in Indiana)

The Warrington Group Ltd.
Professional Products Division
7 Greenleaf Woods Dr.
Portsmouth, NH 03801
603/431-1515
800/662-3338
(fax) 603/431-3232

Weinbrenner Shoe Co., Inc.
108 S. Polk St.
Merrill, WI 54452
715/536-5521
800/826-0002

Wells Lamont Industrial Products
7525 N. Oak Park Ave.
Niles, IL 60714
800/247-3295
(fax) 708/647-8301

Werner Works, Inc.
P.O. Box 974
1931 N.W. Mullholland Dr.
Roseburg, Oregon 97470
800/547-0976
(fax) 503/673-4793

Workrite Uniform Co.
500 E. Third St.
P.O. Box 1192
Oxnard, CA 93032
805/483-0175
800/521-1888

Resources

Videotapes and films:

"Intent vs. Impact." Sexual harassment prevention videotape. Available from Anderson Davis, BNA Communications, Inc., 9439 Key West Ave., Rockville, MD 20850.

"Meeting the Challenge." 10-minute recruitment videotape for women firefighter candidates. Available from Women in the Fire Service, Inc.; P.O. Box 5446 Madison, WI 53705.

"Sex, Power & the Workplace." 60-minue videotape on sexual harassment, with accompanying resource booklet. Available from Lifeguides/KCET Video; 4401 Sunset Boulevard; Los Angeles, CA 90027; 800/343-4727.

"The Job Interview." 11-minute videotape presenting fire department gender stereotypes in a humorous (role-reversal) scenario. Available from Women in the Fire Service, Inc.; P.O. Box 5446, Madison, WI 53705.

"Trade Secrets: Blue Collar Women Speak Out." 23-minute film on women working in the trades. Available on loan from Chicago Women in Trades, 37 S. Ashland St., Chicago, IL 60607; 312/942-1444.

"Valuing Diversity." 5-part videotape series. Available from Copeland and Griggs Prudctions, 302 23rd Ave., San Francisco, CA 94121. Includes titles such as "Managing Differences," "Diversity at Work," "Communicating Across Cultures," etc.

"What About You?/Á toi de choisir!" 19 minute videotape profiling six women working in non-traditional occupations, including a firefighter. Available from Women's Bureau, Labour Canada, Ottawa, Ontario K1A 0J2, Canada; 819/953-0055.

Books and other print resources:

Chetkovich, Carol. *Real Heat: Gender and Race in the Urban Fire Service.* Rutgers University Press, 1997.

Devlin and Associates. *Employment Equity Reference Manual for Ontario Municipal Fire Departments.* Prepared for the Office of the Fire Marshal, Ministry of the Solicitor General, 1991.

FEMA/USFA. *Health and Safety Issues of the Female Emergency Responder.* 1996.

_____. *Physical Fitness Coordinator's Manual for Fire Departments.* 1990.

_____. *Stress Management: Model Program for Maintaining Firefighter Well-Being.* 1991.

Loden, Marilyn. *Workforce America! Managing Employee Diversity as a Vital Resource.* Business One Irwin, 1991.

MacKinnon, Catherine A. *Sexual Harassment of Working Women.* Yale University Press, 1979.

Martin, Molly, ed. *Hard-Hatted Women.* Seal Press, 1989.

Petrocelli, William, and B.K. Repo. *Sexual Harassment on the Job.* Nolo Press, 1992.

Sanders, Jo Schuchat. *The Nuts and Bolts of NTO: How to Help Women Enter Non-Traditional Occupations.* Scarecrow Press, 1986.

Simons, George, and D. Weissman. *Men and Women: Partners at Work.* Crisp Publications, 1990.

Tannen, Deborah. *You Just Don't Understand: Women and Men in Conversation.* Random House, 1990.

_____. *Talking From 9 to 5.* William Morrow, 1994.

Women's Bureau. *Work and Family Resource Kit.* U.S. Department of Labor, 200 Constitution Ave., NW, Room S-331, Washington, D.C. 20210. (Single copies available free.)

Women's Issues Advisory Committee. *Guidelines for Integration of Women into the California Fire Service.* California Fire Fighter Joint Apprenticeship Program, 1990.

Organizations and their publications:

International Association of Black Professional Fire Fighters, 8700 Central Avenue; Landover, MD 20785; 301/808-0804. The IABPFF has national, regional, and chapter committees on Black Women in the Fire Service.

International Association of Fire Fighters, 1750 New York Ave., NW, Washington, DC 20006; 202/737-8484. The IAFF makes available to its members the IAFF Manual on Human Relations, a "Hair Kit" on fire department grooming standards, the original IAFF/USFA manual, *Managing the Entry of Women in the Fire Service,* and a synopsis of information on reproductive safety, pregnancy, and collective bargaining.

National Association of Hispanic Firefighters, 8035 East R.L. Thornton Freeway, Suite 106; Dallas, TX 75228; phone 214/327-8161; E-mail: Tnahfnp@aol.comT The NAHF has chapters in many States.

9 to 5, National Association of Working Women, 614 Superior Ave., NW; Cleveland OH 44113; 216/566-9308. Job Problem Hotline: 800/522-0925 (from Ohio: 216/621-9449). Resources and guidance for women on sexual harassment and other work-related concerns.

NOW Legal Defense and Education Fund, 99 Hudson St., New York NY 10013; 212/925-6635. Information on sexual harassment; guidance on antiharassment policy development.

Women in the Fire Service, Inc., P.O. Box 5446, Madison, WI 53705; 608/233-4768; fax 608/233-4879; E-mail: Tinfo@wfsi.orT. WFS publishes a quarterly periodical on fire service women's issues, women firefighters' recruitment literature, and information packets on a wide range of issues. WFS also holds conferences on fire service women's issues and a biennial leadership training seminar.

Women's Legal Defense Fund, 1875 Connecticut Ave. NW, Suite 710; Washington, DC 20009; 202/986-2600. Information on sexual harassment; advocacy on harassment and other sex discrimination issues.

Appendix: legal issues

Sex discrimination: general

What forms of discrimination are illegal? Discrimination based on race, sex, religion, national origin, ancestry, disability, and, in some States or localities, sexual orientation, political affiliation, marital status, and arrest record. Discrimination is prohibited in all terms, benefits, and conditions of employment, including hiring, firing, layoff, promotion, wages and compensation, fringe benefits, assignment, and training.

Applicable laws:

Title VII of the Civil Rights Act of 1964, as amended, 42 U.S.C. §§ 2000e et seq. This law prohibits discrimination in employment on the basis of race, color, religion, national origin, or sex. This discrimination includes sexual harassment. As amended (see below), Title VII protects pregnant women employees from job discrimination. Employers with 15 or more employees are covered by Title VII.

Pregnancy Discrimination Act of 1978 (amending §701 of Title VII), 42 U.S.C. §2000e (K): pregnancy-related conditions are to be treated the same as other disabling conditions.

Equal Pay Act of 1963, 29 U.S.C. §206(d). This law is an amendment to the Fair Labor Standards Act, and requires that men and women doing substantially equal work receive equal pay.

State and Local Fiscal Assistance Act of 1972: "revenue-sharing" for public safety.

Civil Rights Act of 1991. (Also see p. 142.) This act overturned five 1989 Supreme Court decisions that had gone against complainants of employment bias, as well as reversing two 1991 Supreme Court decisions. It allows compensatory and punitive damages in cases of intentional discrimination based on sex, religion, or disability.

Americans with Disabilities Act of 1990, P.L. 101-336. Prohibits workplace discrimination against individuals who have physical or mental handicaps that affect a significant life function. The ADA applies to employers with 15 or more employees, as well as to places of public accommodation and services.

Rehabilitation Act of 1973. Requires government contractors and recipients of Federal financial assistance to ensure nondiscrimination in employment against disabled employees and applicants.

Executive Order 11246 (1965), as amended: prohibits employment discrimination on the basis of race, color, national origin, or sex in institutions or agencies with Federal contracts over $10,000 (including "grants" that involve a benefit to the Federal government).

Intergovernmental Personnel Act of 1970: Agencies or programs of State and local governments that receive grants-in aid from the Federal government.

Civil Rights Act of 1871, 42 U.S.C. §1983: creates no statutory rights in and of itself, but has been used to enforce federally protected rights from other sources (not including Title VII).

Age Discrimination in Employment Act of 1967, 29 U.S.C. §621 et seq.

The volunteer fire service and gender-based discrimination

A woman volunteer firefighter seeking to obtain the protections of Title VII would encounter the limitation that Title VII applies only to "employees." Volunteers may have legal remedies other than Title VII, but being considered as an employee will significantly increase the volunteer's protection against sex discrimination. Courts have consistently ruled that under Title VII, volunteers are not employees.[1]

In many States, however, volunteer firefighters are considered to be employees for the purpose of anti-discrimination law. This determination may depend on many factors, including how State fair employment practices laws are worded, whether volunteers receive any monetary compensation for their services, and whether they are part of a State pension system. Even if the volunteer is not considered to be an employee, State tort actions would still be available, i.e., assault and battery, intentional infliction of emotional distress, etc. If not covered under Title VII, volunteers still could be protected by the Civil Rights Act of 1871.

[1] *Tadros v. Coleman*, 898 F.2d 10 (2d Cir. 1990), *Smith v. Berks Community Television*, 657 F. Supp. 794 (E.D. Pa. 1987). The Tadros court said plaintiff could only be an employee if the defendant both controls his/her work and pays him/her.
[2] 42 U.S.C. § 1983.

Reproductive issues

Two U.S. Supreme Court decisions, *Johnson Controls* and *California Federal Savings & Loan*, have significant impact on maternity leave issues and the rights of fertile and/or pregnant workers.

UAW v. Johnson Controls 111 S. Ct. 1196 (1991)

The EEOC's Policy Guidance states: "As a result of the Supreme Court's decision in *Johnson Controls*, policies that exclude members of one sex from a workplace for the purpose of protecting fetuses cannot be justified under Title VII. Thus, if a charging party alleges that the respondent has excluded members of one sex from employment based on a fetal protection policy, and if the investigation confirms this allegation, 'cause' should be found. It does not matter whether the employer can prove that a substance to which its workers are exposed will endanger the health of a fetus. It also does not matter whether the employer can prove that it will incur a higher cost as a result of hiring women. Individuals who can perform the essential functions of a job must be considered eligible for employment, regardless of the presence of workplace hazards to fetuses." C.C.H. Empl. Prac. Guide ¶5306 (6/28/91).

The Court's decision: Sex-specific fetal protection policies are, on their face, sex discrimination under Title VII because such policies require only female employees to produce proof that they are not capable of reproducing. Policies which exclude employees on the basis of gender and childbearing capacity rather than fertility are also sex discrimination under the Pregnancy Discrimination Act. A policy that discriminates on the basis of sex can be defended only if it is a "bona fide occupational qualification" (BFOQ). BFOQ must be defined as skills and aptitudes that affect the employee's ability to do the job. "The BFOQ is not so broad that it transforms this deep social concern (the possibility of injury to future children) into an essential aspect of battery-making." "Decisions about the welfare of future children must be left to the parents who conceive, bear, support and raise them rather than to employers who hire those parents."

Maternity policies that require a woman to leave fire suppression duty at a specified point in her pregnancy no longer will be considered legal. However, a fire department administration that offers its firefighters both education regarding the potential reproductive hazards of firefighting and meaningful alternative duty while the employee is attempting to conceive a child or is pregnant, will be offering its employees the option of having a child and having a job.

California Federal Savings & Loan v. Guerra, 107 S. Ct. 683, 42 F.E.P. 1073 (1987)

The Court's decision: "Congress intended the Pregnancy Discrimination Act to be a 'floor beneath which pregnancy disability benefits may not drop--not a ceiling above which they may not rise.'" (at 692)

Antinepotism rules

Antinepotism rules usually are regarded by fact-finders as "sex-neutral" in that on their face they apply to both men and women equally. A complainant must present ample statistics to show disparate impact* and that the rule is not justified by business necessity. A complainant also must check to see if State and local Fair Employment Practices (FEP) laws would (1) ban discrimination on the basis of marital status, and/or (2) enable the no-spouse rule to be defended as BFOQ.

Relevant cases:

Yuhas v. Libby-Owens-Ford, 562 F.2d 496 (7th Cir. 1977), *cert denied,* 435 U.S. 934 (1978), *rev'g* 411 F. Supp. 77 (N.D. Ill. 1976).

Plaintiff's statistics established prima facie case of sex discrimination but court accepted employer's "business" justifications that employment of both spouses in the same capacity would hurt the morale of other employees and of the spouses themselves.

EEOC Decision No. 75-239 (1976), CCH EEOC Decisions ¶ 6492.

Employer did not meet business necessity burden; e.g., previous history had allowed relatives to work and there was no evidence the employer had had problems before instituting the antinepotism rule.

Volchahoske v. City of Grand Island, 10 E.P.D. ¶10,247 (Neb. 1975).

Ordinance prohibiting employment of both husband and wife found unconstitutional interference with "right to marry" protected under the 1st, 5th, 9th, and 14th Amendments [city must show compelling interest].

Sebetic v. Hagerty, 640 F. Supp. 1274 (E.D.Wis. 1986), *aff'd* 819 F.2d 1144 (7th Cir.), *cert denied* 108 S. Ct. 235 (1987).

Rule prohibited spouses of law enforcement officers from being dispatchers. Court held no-spouse policy valid based on need to avoid dangerous incidents that may occur when dispatcher's spouse is officer on call. The rule did not significantly interfere with the right to marry, and the policy was not a pretext for sex discrimination.

Espinoza v. Thoma, 580 F.2d 346 (8th Cir. 1978)

Court applied a no-spouse rule to an unmarried couple.

---

*If the rule is "no spouse" rather than "no relative," plaintiff must establish the proposition that it is most often the wife who is terminated under the "no spouse" rule.

The Civil Rights Act of 1991

The Civil Rights Act of 1991[1] had as its principal thrust the reversal of a number of recent Supreme Court decisions (many of which were cited and discussed in the first edition of this *Handbook*). Arguably, it was the public's heightened awareness of the implications of sexual harassment, due to publicity over the Hill-Thomas hearings and the Tailhook scandal, that propelled Congress and the President to approve legislation that was substantially similar to a bill vetoed the year before.

The Act resulted in the following changes in the area of sex discrimination litigation:[2]

1. Compensatory[3] and punitive damages can be awarded where a plaintiff establishes that intentional sex discrimination occurred. Such damages are expressly **not** available in disparate-impact cases.

2. Damages are awarded on a sliding scale, with the upper limits based on the size of the employer's business.

3. Plaintiffs have the option of asking for a jury trial.

4. The Supreme Court's antiplaintiff definition of "business necessity" in *Wards Cove Packing Co., Inc. v. Atonio*[4] is overruled. This had allowed employers to offer any generalized business objective as a justification for the necessity for an action resulting in adverse impact on a protected group, rather than having to prove the job-relatedness of the discriminatory practice.

5. Employment-related test score adjustment based on race, color, sex, religion, or national origin is prohibited.

6. Stricter limits govern after-the-fact challenges to litigated judgments and consent decrees in discrimination cases.

7. The application of Title VII to employers operating outside the U.S. is expanded and clarified.

8. The right to challenge discriminatory seniority systems is expanded.

9. The statute of limitations for filing lawsuits against the Federal government as an employer is extended to 90 days.

10. Expert fees may be awarded with attorneys' fees.

11. Prejudgment interest is awardable in Federal employment Title VII lawsuits.

12. A "Glass Ceiling Commission" (studying upper-management barriers to women) was created.

13. A technical assistance training institute was created to assist employers with compliance with equal employment opportunity laws.

14. Parties are encouraged to use alternative means of dispute resolution (other than the courts).

Since the passage of the Act, the lower courts have been struggling with whether its provisions are applicable to acts and cases occurring before its passage. In one case, the court allowed the plaintiff to take advantage of the Act's jury-trial and punitive-damage provisions where the discrimination was alleged to have been ongoing and continuous after the Act's passage.[5] But in *Landgraf v. U.S. Film Products*, the Supreme Court ruled that the provisions of the Act were not retroactive and the plaintiff should not have been granted a jury trial or compensatory and punitive damages. The entire case was returned to the lower courts for a retrial before a judge.[6] The importance of *Landgraf* is diminishing as we move further in time from 1991.

Notes:

[1] 42 U.S.C. §1981 A.

[2] In some cases, the law "reverted" to definitions and interpretations of Title VII that had been in existence prior to a series of very restrictive and antiplaintiff Supreme Court cases decided in the late 1980's.

[3] "Compensatory" damages are defined by the Act to include not only back pay (which always had been available) but also mental anguish, loss of enjoyment, future pecuniary gains, etc.

[4] *Wards Cove Packing Co. Inc., v. Atonio*, 490 U.S. 642 (1989).

[5] *Russell v. City of Overland Police Dept.*, 838 F. Supp. 1350 (E.D. Mo. 1993).

[6] 114 S. Ct. 1483 (1994).

Bibliography

Affirmative action

"DCFD Affirmative Action Hiring Overturned." *Fire Chief*, May 1987, p. 28.

"Female Firefighters Decline Affirmative Action Promotions." *Fire Engineering*, 145(5) (May 1992): p. 16.

Osby, Robert E. "Guidelines for Effective Fire Service Affirmative Action." *Fire Chief*, Sept. 1991, pp. 50-54.

Schumacher, Joe. "Affirmative Action Revisited." *Fire Chief*, Mar. 1989, pp. 51-53.

Slack, James D. "Women, Minorities, and Public Employer Attitudes: The Case of Fire Chiefs and Affirmative Action." *Public Administration Quarterly*, 13(3) (Fall 1989): pp. 388-411.

Family issues

"London Fire Brigrade Considers Child Care Alternatives." *IAFC On Scene* (Sept. 15, 1991): pp. 1-2. [Condensed from an article in *Women in the Fire Service Quarterly*, Summer 1991.]

Willing, Linda. "Love on the Job." *Fire Chief*, Aug. 1990, p. 92.

Firefighter training and pretraining programs

Bird, James W. "Training Women for the P.A.T." *Fire Engineering*, Mar. 1991, pp. 87-93.

"Eleven Women Survive NYFD Training Program." *International Fire Chief*, 49(8) (Jan. 1983): p. 8.

Larkin, Susan R. "Training for Success." *Fire Command*, Aug. 1989, pp. 38-42. [Training women for FDNY entry-level testing.]

McDonald, Bernie R. "Pre-Employment Training: One Department's Program." *Fire Command*, August 1987, pp. 24-28. [Louisville Fire Department.]

Pletan, Owen D. "A Format for Successful Pre-recruit Training: The Seattle Method." *Fire Command*, 48(8) (Aug. 1981), pp. 35-37.

"Women's Training Program in Jacksonville." *Fire Chief*, Feb. 1991, pp. 60-61.

"Women's Training Program Upgrades Firefighting Skills." *Fire Chief*, Feb, 1991, pp. 60-61.

Legal issues

"Department of Justice Consent Decree." *Fire Chief*, July 1987, p. 30.

"Newest Court Ruling Backs Female Hires." *International Fire Fighter*, 70(5-6) (May-June 1987): p. 1.

Rukavina, John. "Fire Service, Meet the ADA." *Fire Chief*, June 1992, pp. 30+. [Civil Rights Act of 1991.]

_____. "Seeing the Future." *Fire Chief*, Sept. 1992, pp. 22+. [Regarding Pennsylvania State University report on the pending expiration of the Age Discrimination in Employment Act exemption for firefighters.]

Schmidt, Wayne W. "Background Investigations." *Fire Chief*, July 1990, p. 18. [Limits on background investigations in areas such as sexual activity.]

_____. "Civil Liability for Wrongful Discharge," *Fire Chief*, Sept.1989, pp. 26+.

_____. "Nepotism and Consanguinity Relations." *Fire Chief*, Feb. 1985, p. 14. [9th Circuit affirms lower court ruling that one employee may be required to transfer or quit if two employees marry.]

Shearer, Robert W. "Can After-Hours Conduct Be Grounds for Firing?" *Fire Chief*, May 1989, pp. 59-60.

"US Court of Appeals Throws Out FDNY Scoring System." *Fire Chief*, May 1987, pp. 14-15.

Physical fitness and physical performance testing

Bell, Laura. "Where Does Physical Testing Leave Women?" *Management Review*, Dec. 1987, pp. 47-50.

Clark, Allen. "Female Firefighters Not the Issue--It's Physical Fitness." *Fire Chief*, Oct. 1991, p. 22.

Davis, James E. "A Look at Performance Standards." *Fire Chief*, Aug. 1991, p. 56.

Doolittle, T.L. "Validation of Physical Requirements for Firefighters." Seattle Fire Department, Office of Management and Budget, Seattle, 1979.

Evans, D.H. "Height, Weight and Physical Agility Requirements." *Journal of Police Science and Administration*, Dec. 1980, pp. 414-436.

FEMA/USFA, *Physical Fitness Coordinator's Manual for Fire Departments*, 1990.

George, Arthur E. "Only One Standard." *Fire Engineering*, 141(3), pp. 37-38.

Kay, Herbert. "Testing Recruits." *Fire Chief*, Apr. 1989, pp. 70+.

Misner, J.E., S.A. Plowman, and R.A. Bioleau. "Performance Differences between Males and Females on Simulated Firefighting Tasks," *Journal of Occupational Medicine*, 29, (1987), pp. 801-805.

Rafilson, Fred M. "Legislative Impact on Fire Service Physical Fitness Testing." *Fire Engineering*, 148(4) (Apr. 1995): pp. 83-84.

Schmidt, Wayne, W. "Physical Fitness Standards." *Fire Chief*, July 1989, p. 42. [Legality of employment requirements restricting body fat.]

Williams, Timothy, and S. Evenson. "Physically Fit For Duty? By Whose Standards?" *Fire Chief*, Mar. 1988, pp. 43+; Apr. 1988, pp. 58+; May 1988, pp. 55+.

Promotion

Enbysk, Liz Peeples. "First Female Fire Chief in Paid Position." *American Fire Journal*, 35,(10) (Oct. 1983): p. 14.

Glass Ceiling Commission. *A Solid Investment: Making Full Use of the Nation's Human Capital: Recommendations of the Federal Glass Ceiling Commission.* Washington, D.C., Nov. 1995.

_____. *Good For Business: Making Full Use of the Nation's Human Capital; The Environmental Scan.* Washington, D.C., Mar. 1995.

Hirschman, Jessica. "Climbing the Glass Ladder--Part II. *Firefighter's News*, 11(3) (June-July 1993): pp. 44-47.

Protective clothing

Duffy, Richard; J. Sawicki, and A. Beer. "Project FIRES: Final Report." International Association of Fire Fighters, 1985.

National Fire Protection Association. "NFPA 1971: Standard on Protective Clothing for Structural Fire Fighting." 1991.

_____. "NFPA 1972: Standard on Helmets for Structural Fire Fighting." 1987.

_____. "NFPA 1973: Standard on Gloves for Structural Fire Fighting." 1988.

_____. "NFPA 1974: Standard on Protective Footwear for Structural Fire Fighting." 1987.

_____. "NFPA 1975: Standard on Station/Work Uniforms for Structural Fire Fighters." 1990.

_____. "NFPA 1981: Standard on Open-Circuit Self-Contained Breathing Apparatus for Structural Fire Fighting." 1987.

Neeves, R., *et al.*, "Physiological and Biomechanical Changes in Fire Fighters Due to Boot Design Modifications." International Association of Fire Fighters and the Federal Emergency Management Agency, 1989.

Sylvia, Dick. "New Turnout Gear, Women's Roles Among Topics at IAFC Conference." *Fire Engineering*, 131(11) (Nov. 1978), pp. 58-62.

Recruitment and retention

Bifano, Angie. "Firefighter Selection Today: The Problem of Balancing Legal, Social and Occupational Safety Issues." *Firefighter's News*, 9(5) (Aug.-Sept. 1991): pp. 48-50.

Booth, Walter. "Recruiting Women and Minorities." *Fire Chief*, May 1987, pp. 49-53. [Survey of large fire departments regarding their recruitment strategies.]

Brown, Marsha D. "Getting and Keeping Women in Nontraditional Careers." *Public Personnel Management Journal*, Winter 1981, pp. 408-411.

Durkin, Edward D. "Recruiting and Hiring Women Firefighters." *Fire Chief*, May 1981, pp. 52-55.

Goldfeder, William. "Retaining and Recruiting Members." *Fire Engineering*, May 1992, pp. 10-13.

Hammond, Ken. "Recruiting Women Firefighters." *Fire Chief*, Oct. 1987, pp. 40-41.

Makela, William. "Women Taking Up the Slack in Recruiting." *Minnesota Fire Chief*, 27(1) (Sept.-Oct. 1990): pp. 12-13, 59.

Marinucci, Richard A. "Attracting Recruits: A Matter of Image." *Fire Engineering*, July 1991, p. 10. [Recruiting volunteer firefighters.]

Sanders, Jo Schuchat. *The Nuts and Bolts of NTO: How to Help Women Enter Non-Traditional Occupations*. Scarecrow Press, 1986.

Scotford, Garth. "Programs Revealed Problems in Attracting Women Recruits." *Fire*, 82(1019) (May 1990): p. 9. [Letter to the editor.]

Thaut, Stanley L. "A History of Tacoma's Effort to Recruit Women Firefighters." *Fire Chief*, 23(9) (Sept. 1979): pp. 40-43.

Waters, Michael S. "The Recruitment and Retention of Women in the Career Fire Service." *International Fire Chief*, May 1986, pp. 14-17.

_____. "The Recruitment and Retention of Women in the Career Fire Service, Part II." *International Fire Chief*, June 1986, pp. 18-21.

_____. "The Recruitment and Retention of Women in the Career Fire Service, Part III." *International Fire Chief*. July 1986, pp. 20-23.

Reproductive Safety

Dixon, Ernest M. "Reproductive Disorders of the Male/Female Worker: Occupational Placement of Women of Reproductive Capacity--Views of the Medical Community." Occupational Safety and Health Symposia, 1978, pp. 27-28. Cincinnati, Ohio: U.S. National Institute for Occupational Safety and Health, June 1979.

Evanoff, Bradley A., and Linda Rosenstock. "Reproductive Hazards in the Workplace: A Case Study of Women Firefighters." *American Journal of Industrial Medicine*, 9(6) (1986): pp. 503-515.

Fischer, David R., William A. Jones, Charles A. Lacroix, Clay A. Phillips, Perry E. Ray, and Timothy P. Travers. *Written Policies and Pregnant Firefighters*. Emmitsburg, MD: National Fire Academy, June 10-21, 1991.

Infante, Peter F. "Reproductive Disorders of the Male/Female Worker: Occupational Placement of Women of Reproductive Capacity--OSHA's View." Occupational Safety and Health Symposia, 1978, pp. 29-30. Cincinnati: U.S. National Institute for Occupational Safety and Health, June 1979.

McDiarmid, Melissa, M.D., *et al*. "Reproductive Hazards of Firefighting I and II." *American Journal of Industrial Medicine*, 19:433-472 (1991).

Olshan, Andrew F., K. Teschke, and P. Baird. "Birth Defects Among Offspring of Firemen." *American Journal of Epidemiology*, 131(2) pp. 312-321.

Stellman, Jeanne M. "Reproductive Disorders of the Male/Female Worker: The Effects of Toxic Agents on Reproduction." Occupational Safety and Health Symposia, 1978, pp. 16-26, Cincinnati, Ohio: U.S. National Institute for Occupational Safety and Health, June 1979.

Templeton, Randy. "Pregnant Firefighter." *Fire Chief*, 36(4) (Apr. 1992): pp. 116-118, 120.

Sex discrimination/Sexual harassment

Barron, Donna. "Men, Women and Harassment." *Emergency*, 20(7) (July 1988): pp. 31-35.

Blackistone, Steve. "Sex Discrimination." *Firehouse*, 15(11) (Nov, 1990): p. 69.

_____. "Sexual Harassment in the Firehouse (Fire Law column)." *Firehouse*, 17(4) (Apr. 1992): pp. 90-92.

Beekman, Peter and Allen Fankhauser. "Like It or Not, Sex Is Here to Stay." *JEMS*, 16(1) (Jan. 1991): pp. 13-14.

"Cleveland FD Charged With Discrimination." *International Fire Chief*, 50(2) (Feb. 1984): p. 10.

Farley, Lin. *Sexual Shakedown: The Sexual Harassment of Women on the Job*. McGraw-Hill, 1978.

"Firefighters Transferred in Queens Sex-Bias Case." *Fire Control Digest*, 17(11) (Nov. 1991): pp. 6-7.

"Justice Actions." *Fire Chief*, Nov. 1986, p. 25. [Charleston, WV, Police Department ordered to reinstate dispatcher fired when she became pregnant.]

"Justice Department Acts in Discrimination Cases." *Fire Chief*, July 1985, pp. 8-10. [Delete numerical hiring goals implemented in 1977 consent decree in San Diego.]

"Justice Department Sues City." *Fire Chief*, November 1991, p. 38. [Dept. of Justice alleges discrimination against white men when West Palm Beach directed the hiring of minorities and women to fill twelve vacancies.]

MacKinnon, Catherine A. *Sexual Harassment of Working Women*. Yale University Press, 1979.

McQueen, Iris. "Sexual Harassment." *Fire Chief*, Aug. 1985, pp. 69-72.

Moore, Thomas V. "Sexual Harassment in the Firehouse?" *Firehouse*. 10(18) (Aug. 1985): p. 14.

Naczi, Frances D. "Removing Sexism From Communications." *Fire Chief*, Nov. 1984, pp. 45-46.

Paludi, Michele A., and R.B. Barickman. *Academic and Workplace Sexual Harassment*. State Univ. of New York, 1991.

Petrocelli, William, and B.K. Repa. *Sexual Harassment on the Job*. Nolo Press, 1992.

Randleman, William. "What is Discrimination?" *Fire Chief*, Nov. 1984, p. 27.

Schmidt, Wayne W. "Sex Discrimination." *Fire Chief*, Apr. 1982, p. 16.

_____. "Quotas and Other Discrimination Remedies." *Fire Chief*, Apr. 1985, p. 14.

_____. "Sexual Harassment." *Fire Chief*, Oct. 1986, pp. 14-15. [U.S. Supreme Court decision in *Meritor v. Vinson*; also Indiana case where judge found sexual harassment but not sex discrimination.]

_____. "Sexual Harassment." *Fire Chief*, Feb. 1987, p. 27. [Appeals court reversal of lower court decision that failed to find harassment evidence of discrimination.]

_____. "Sexual Harassment." *Fire Chief*, Dec. 1991, p. 30. [Cases involving women police officers.]

Schrader, George. "Avoid Sexual Harassment Hassles." *Fire Chief*, June 1990, pp. 47+. [Sexual harassment not an issue of sex but of power and control; recent court cases.]

"Sex Discrimination." *Fire and Police Personnel Reporter*, Aug. 1984, pp. 13-15.

"Sexual Harassment." *Fire and Police Personnel Reporter*, Oct. 1986, pp. 12-14.

Shouldis, William. "Sexual Harassment." *Fire Engineering*, Sept. 1991, pp. 101+.

Siegel, Deborah L. *Sexual Harassment: Research and Resources*. National Council for Research on Women, 1991.

Webster, Cindy. "Facing Off On Sexual Harassment." *Fire Chief*, Aug. 1992, pp. 72-77.

Supporting workforce diversity

Belenky, Mary Field, *et al.*; *Women's Ways of Knowing*, Basic Books, 1986.

Bell, Derrick, *Faces at the Bottom of the Well*. Basic Books, 1992.

Brightmire, Susan. "Diversity." *Fire Engineering*, 148(1) (Jan. 1995): p. 117-118.

Briese, Gary. "The Challenge of the '90s: Prepare for the 21st Century." *The Minnesota Fire Chief*, 28(4) (Mar./Apr. 1992): pp. 14-15.

Cannon, Katie G., *Black Womanist Ethics*, Scholars Press, 1988.

Cary, Lorene, *Black Ice*, Knopf (Random House), 1991.

Gilligan, Carol, *In a Different Voice*, Harvard University Press, 1982.

Helgesen, Sally, *The Female Advantage*, Doubleday, 1990.

Johnston, William B. *Workforce 2000: Work and Workers for the 21st Century*. Hudson Institute, 1987.

Katz, Judith H., *White Awareness: Handbook for Anti-Racism Training*, University of Oklahoma Press, 1978.

Kegan, Robert, *In Over Our Heads*, Harvard University Press, 1994.

Kochman, Thomas, *Black & White: Styles in Conflict*, University of Chicago Press, 1981.

Lightfoot, Sara Lawrence, *Balm in Gilead*, Addison-Wesley Pub. Co., 1988.

_____., *I've Known Rivers*, Addison-Wesley Pub. Co., 1994.

Loden, Marilyn, *Workforce America! Managing Employee Diversity as a Vital Resource*, Business One Irwin, 1991.

Macklin, Victoria S. "Peer Mediation Helps Heal a House Divded." *NJPA Journal*, May-June 1991, pp. 62-66.

Morrison, Ann M., *Breaking the Glass Ceiling*, Addison-Wesley Pub. Co., 1992.

_____., *The New Leaders: Guidelines on Leadership Diversity in American*, Jossey-Bass, 1992.

Perry, Linda A.M., Turner, Lynn H. & Sterk, Helen M. (eds.); *Constructing and Reconstructing Gender*, SUNY Press, 1992.

Simons, George. *Working Together. How to Become More Effective in a Multicultural Organization*. Crisp Publications, 1989.

Simons, George, and D. Weissman. *Men and Women: Partners at Work*. Crisp Publications, 1990.

Smith, Michael H. "Communications Skills for a Changing Fire Service." *Fire Chief*, Sept. 1991, p. 82.

Sturzenacker, Gloria. "Prejudice Prevention." *Chief Fire Executive*, Apr.-May 1986, pp. 43-51.

Takaki, Ronald T., *A Different Mirror*, Little, Brown & Co., 1993.

Tannen, Deborah, *Gender and Discourse*. Oxford University Press, 1994.

_____. *Talking from 9 to 5*. William Morrow, 1994.

_____. *You Just Don't Understand: Women and Men in Conversation.* Ballantine Books, 1990.

Three Rivers, Umoja; *Cultural Etiquette,* Market Wimmin, 1990.

Volunteer firefighters

Cashman, John R. "Gal Fire Fighters Do What's Needed in Brigade Too Small for Specialists." *Fire Engineering,* 132(12) (Dec. 1979), pp. 14-16.

Chambers, Mary D. "Volunteer Fire Chiefing." *International Fire Chief,* Aug. 1980, p. 15.

Marinucci, Richard A. "Women in the Volunteer Fire Service." *Fire Engineering,* Jan. 1991, pp. 10-12.

Mitchell, Carol Ann. "A History of Service: Women in Volunteer Fire Companies." *California Fire Service,* 2(6) (June 1991): p. 24.

Perkins, Kenneth B. "Volunteer Fire Fighters in the U.S: A Sociological Profile of America's Bravest." National Volunteer Fire Council, 1987.

Women firefighters

"All Female Fire Brigade." *Rekindle.* 14(9) (Sept. 1995): p. 5.

Beaver, Don R., and Jerry Knapp. "Women in the Fire Service." *Fire Command,* 51(8) p. 15.

Casey, Jim. "Women Fire Fighters Are Here to Stay." *Fire Engineering,* 131(3) (Mar. 1978), p. 6.

Chambers, Mary D. "Women in the Fire Service." *Western Fire Journal,* Apr. 1981, p. 40.

Chapman, Brenda J. "Women in the Fire Service: Do We Belong?" *Voice,* 20(5) (May 1991): p. 11

Clayton, Bill. "Female Inmates Work as Wildland Firefighters." *American Fire Journal,* 36(10) (Oct. 1984): 45-46.

Cridland, Elizabeth. "Ideas for Increasing Women's Role in Fire Service Voiced at Seminar." *Fire Engineering,* 134(4) (Apr. 1980): pp. 28, 30.

DeMars, Denise. "What's the Big Deal? (Thoughts of an Individual Fire Fighter)." *Fire Command,* 51(8) (Aug. 1984): p. 23.

Dernocoeur, Kate, EMT-P, and James N. Eastman, Jr., ScD. "Have We Really Come a Long Way? Women in EMS Survey Results." JEMS, 17(2) (Feb. 1992): pp. 18-19.

Dessoff, Alan L. "Female Inmates: No-Holds-Barred Brigade." *Firehouse,* 48(8) (Aug. 1981): pp. 35-37.

Feldman, Danah. "Wildland Fire Fighting." *International Fire Chief,* 46(8) (Aug. 1980), pp. 16-17.

"Fire Division Gets First Female Apparatus Operator." United Press International, Regional News--Cincinnati, OH, (Jan. 6, 1992).

Floren, Terese M. "1990 Survey Results." *WFS Quarterly,* Winter 1990-1991, pp. 14-17.

_____. "Women Firefighters: The Chief's Role." *Fire Chief,* May 1981, pp. 48-51.

_____. "Women Firefighters Speak: A Survey of the Nation's Female Firefighters." *Fire Command*, Dec. 1980, pp. 22-24, Jan. 1981, pp. 22-25.

Granito, Dolores. "More Women Entering Fire Service." *Fire Engineering*, 131 (Mar. 1978), pp. 29-30.

Hallinan, Lorin. "Breaking the Barriers." *Emergency*, 26(5) (May 1994), pp. 32-37.

Hamilton, Jo Carol. "Women in the Fire Service." *Fire Chief*, 22(8) (Aug. 1978), pp. 81-84.

Istvan, Sharon. "Fire Protection Engineering." *International Fire Chief*, Aug. 1980, p. 21.

Keene, Kathy. "What Is It Like To Be a Female Firefighter?" *Fire Chief*, Sept. 1991, pp. 72-74.

Keller, D.F. "Women Prisoners Protect Facility, Serve Outside Area." *Fire Engineering*, 132(8) (Aug. 1979), p. 124.

Knapp, Jerry and Don R. Beaver. "Women in the Fire Service: Two Views." *Fire Command*, (8) (Aug. 1984): p. 15.

Lipkin, Harriett. "Smoothing the Way for Women." *International Fire Chief*, Aug. 1980, pp. 22-24.

Love, Myron. "Still a Tough Job, But No Longer a 'Man's World'. "*Fire Fighting in Canada*, 36(2) (Mar. 1992): pp. 6-7.

"Niche Jobs Can Be Found for Women." *Fire*, 82(1018) (Apr. 1990): p. 25.

"Oregon Volunteer Firefighter of the Year: Captain Mary Lou Fletcher." *Fire Command*, Sept. 1988, p. 6.

Orr, Robert. "Women's Work?" *Fire Prevention*, (244) (Nov. 1991): p. 10.

Pantoga, Fritzie. "Women Firefighters--A Survey." *Fire Chief*, Jan. 1977, pp. 51-54.

Perkins, Kenneth B. "Women in the Ranks." *Firehouse*, 8(3) (Mar. 1983), pp. 49-50.

Quinn, Richard C. "First Women Fire Fighters." *Fire Command*, 47(9) (Sept. 1980), p. 29.

Roche, Diane C. "Public Fire Education." *International Fire Chief*, 46,(8) Aug. 1980, pp. 20-21.

Rudder, Beatrice. "Career Fire Fighting." *International Fire Chief*, 46(8) Aug. 1980, p. 16.

Rule, Charles H., R.E. Osby, J.H. Steffens, and M.R. Rakestraw. "Workforce 2000." *Fire Chief*, Jan. 1991, pp. 36-40.

Senk, Terry A. "The Firefighter is a Lady--Women in the Fire Service." *The Voice*, 20(4) (Apr. 1991): pp. 8-9.

Sexson, Margarita Y. "Fire Departments Surveyed on Employment of Females." *Fire Engineering*, 134(6) (June 1981): pp. 48-49.

Smith, Dennis. "Women in the Fire Service." *Firehouse*, 8(2) (Feb. 1983), p. 5.

Summers, Liz. "We're Here To Stay: Reflections of a Woman in the Fire Service." *Voice*, 24(5) (June 1995): pp. 41-42.

Swartout, Robert. "Women Fire Fighters: The Seattle Concept." *International Fire Chief*, Aug. 1980, pp. 10-11.

Thompson, F. McKeen. "Woman's Place." *Emergency*, 19(4) (Apr. 1987): p. 4.

Townley, John P. "Hiring Women Fire Fighters Opposed." *Fire Engineering*, 131(3) (Mar. 1978): p. 33.

Walker, Leslie. "A Firefighter Like You." *Voice*, 24(5) (June 1995): pp. 43-44.

Wasylyk, Sylvia. "Women in the Fire Service--Issues and Answers." *Voice*, 23(8) (Sept. 1994), pp. 38-41.

Wauls, Bonita. "The Battle's Over." *Fire Command*, 51(10) (October 1984): p. 6. [Letter to the Editor.]

Webster, Cindy. "First National Conference of Fire Service Women." *Fire Chief*, Feb. 1986, pp. 44-47.

Winkle, William, and R. Navarre. "Females in the Fire Service: The Process of Acceptance." *Fire Chief*, Apr. 1985, pp. 68-69. [How the Toledo Fire Department integrated women into its ranks.]

"Women Loses 9th Try to Join Fire Department." *Fire Control Digest*, 17(1) (Jan. 1991): p. 10.

"Women are Fire Fighters, Too!" *Fire Command*, Feb. 1976, pp. 16-19.

"Women in the Fire Service." *Fire Chief*, May 1981, pp. 48-51.

"Women in the Fire Service. The Challenge in America." *Fire International*, 7(73) (Dec. 1981): pp. 41-43.

"Women Firefighters Still Struggle." *Fire Chief*, 31(3) (Mar. 1987): p. 6.

Women's Issues Advisory Committee. *Guidelines for Integration of Women into the California Fire Service*. California Fire Fighter Joint Apprenticeship Program, 1990.

Ziolkowski, Heidi M. "Vying for a 'Man's Job' Now More Than Wishing on a Star." *Western Fire Journal*, 35(8) (Aug. 1983): p. 5.

Other

Buchbinder, Laura B., and Carol Vougioukles. "Fire Administration Initiates Women's Program." *International Fire Chief*, 46(8) (Aug. 1980), pp. 12-13.

Burton, Mike. "In a Family Sort of Way." *Voice*, 24(5) (June 1995): pp. 17-20.

Chetkovich, Carol. *Real Heat: Race and Gender in the Urban Fire Service*. Rutgers University Press, 1997.

Coleman, Ronny J., and John A. Granito, eds. *Managing Fire Services*. 2nd ed. Washington, DC: International City Management Association, ca. 1988, pp. 230-231; 266-269.

Craig, Jane, and R. Jacobs. "The Effect of Working with Women on Male Attitudes toward Female Firefighters." *Basic and Applied Psychology*, Mar. 1985, pp. 61-74.

Deasy, M. "One Size (Does Not) Fit All." *Firehouse*, May 1988, pp. 33-36.

Devlin & Associates, *Employment Equity Reference Manual for Ontario Municipal Fire Departments*, prepared for the Office of the Fire Marshal, Ministry of the Solicitor General, 1991.

Fire Department Personnel Management Handbook. Managing the Entry of Women and Minorities. Washington, DC: FEMA, 1982.

Guidelines for Integration of Women into the California Fire Service. CA: California Fire Fighter Joint Apprenticeship Committee's Women's Issues Advisory Committee, 1990.

Martin, Molly, ed. *Hard-Hatted Women*. Seal Press, 1989.

McCarl, Robert. *The District of Columbia Fire Fighters Project: A Case Study in Occupational Folklife*. Smithsonian Institute Press, 1985.

McNichol, J., and S. Scanlin. "Proceedings of the National Firefighter Health and Safety Forum." Congressional Fire Services Institute, 1991.

National Fire Protection Association. "NFPA 1500: Standard on Fire Department Occupational Safety and Health Programs." 1987.

Navarre, Raymond J. "Developing a Stress-Reducing Fire Station." *Fire Chief*, Feb. 1987, pp. 46-48.

Paradigm, Inc. *Issues for Women in the Fire Service*. Washington, DC: U.S. Fire Administration, Sept. 1980.

Stress Management: Model Program for Maintaining Fire Fighter Well-Being, Washington, DC: FEMA, 1991.

Thomason, Betsy. "Self-discovery: A Way to Deal With Stress." *Fire Chief*, Feb. 1991, p. 25.

Tokle, Gary. "1001 Considerations." *Fire Command*, Apr. 1989, pp. 24-25.

Turner, Gary. "Butting Heads over Change." *Chief Fire Executive*, Jan.-Feb. 1987, pp. 27-30

Vonada, Michael. "Shadow Dancing." *Fire Chief*, Dec. 1987, p. 50.